To our parents

Aims of this book

This book is an outgrowth of courses in glycobiology given in the final year of the undergraduate biochemistry curricula at Imperial College London and the University of Oxford. It is intended to serve the needs of senior undergraduate students and starting graduate students. However, we hope that it will also provide a useful introduction to the field of glycobiology for more advanced researchers, including biochemists, cell biologists, immunologists, and virologists who encounter glycoproteins, glycolipids, and their receptors during the course of their work. Many people from outside the field see the need to understand what the sugars are doing, but they find the field bewildering. They can easily be put off by the enormous amount of structural detail associated with the sugars and the lack of a clear answer to the question 'Why are these sugars there in the first place?' This book will help educators to present a modern view of glycobiology to students and will provide an overview to practising scientists who are trying to come to grips with the flood of information in the field.

The presentation assumes background knowledge in biochemistry and cell biology at an undergraduate level. Our goal is not to undertake an exhaustive survey of the entire field of glycobiology. Instead, we provide an overview of the functions of glycans and an indication of how these functions are performed at the molecular level. Selective examples are used throughout; many interesting and important stories have, unfortunately, been left out in the interests of keeping the presentation short but representative of the whole field. Specifically, the focus of the book is on glycosylation in mammalian systems. This somewhat arbitrary narrowing of the subject has allowed us to make glycosylation in human disease a theme throughout the book.

Suggestions on how to use this book

Different lecturers will have different views about how to organize the teaching of glycobiology. This book is arranged so that the topics can be covered in essentially any order. The material is presented in relatively brief chapters in which individual sections have been given descriptive headings summarizing the main concepts that are being explained. Key terms are highlighted in the text at the point where they first appear and most are defined in the glossary. The book is divided into two parts. The first part describes the chemical nature of various classes of glycans, their conformations, and their biosynthesis. The second part of the book addresses the biological functions of glycosylation. In this edition, Cross-references have been added in the margins to link topics between chapters, particularly between the two sections.

Within the first section, Chapter 1 provides background on sugar chemistry and oligosaccharide biosynthesis that is essential for most of the topics that follow. In Chapter 2, many of the principles of glycosylation are described in the context of N-linked glycosylation. The roles of different categories of O-linked glycans are described in Chapter 3, with an emphasis on functions that derive from their unique physical properties. This theme is continued in Chapter 4, which is concerned with glycolipids and glycoproteins in membranes. Chapter 5, which is new in this edition, describes the molecular and cellular mechanisms of glycosylation and highlights key outcomes from the study of knockout mice. The material on strategies for analysis of glycans in Chapter 6 and the coverage of glycan conformation in Chapter 7 do not need to be read before the subsequent chapters. These chapters contain material that may be difficult for some students, so many readers will wish to move directly from Chapter 5 to chapters dealing with the biological functions of glycans. Chapters in the second section of the book can be read independently of each other. The ability of glycosylation to affect the behaviour of proteins is discussed in Chapter 8. Chapters 9 and 10 focus on the informational role of glycans and the lectins that decode this information. The roles of lectins in cell adhesion, signalling, and innate immunity are covered in Chapter 9 and lectins that mediate trafficking inside and outside cells are described in Chapter 10.

Because the goal of this book is to provide a concise introduction to glycobiology, the focus is on mammalian systems. However, Chapter 11 provides a glimpse of glycobiology beyond mammalian cells. Chapter 12 focuses on the glycobiology of development. While many of the effects described lack a truly satisfying molecular explanation, this area represents one of the most active in the glycobiology field. The role of glycosylation in disease processes is a continuing theme throughout the book. Marginal notes are provided to highlight points where this subject is discussed and several new boxes devoted to special topics relating to disease and therapeutics have been added to this edition. Further examples are considered in detail in Chapter 13. The final chapter provides some comments on where the field of glycobiology is headed.

Teaching aids are provided with the goal of making the teaching of glycobiology accessible to undergraduates. Learning objectives are presented at the beginning of each chapter. These objectives, as well as the Glossary, are intended to help students to test their understanding of the material. Each chapter and each special topic box ends with a set of Questions and Essay topics. The lists of Key references have been kept brief, with recent review articles cited where possible to provide a broad overview of the field. However, many of the problems and figure legends contain references to research articles to encourage students to explore the primary literature.

 ## Online Resource Centre

The **Online Resource Centre** at **http://www.oxfordtextbooks.co.uk/orc/taylor3e/** contains direct links to the Key references. Updates and additional topics of interest will be added to the site at regular intervals. A list of Protein Data Bank codes is provided to allow the viewing of **3D rotatable structures**. As an aid to instructors, **full colour** versions of the figures are provided for downloading from the website. Suggested answers to the problems are also available to registered adopters of the book. The Online Resource Centre can be accessed by visiting www.oxfordtextbooks.co.uk/orc/taylor3e/.

New to this edition

- Updated content to reflect the current state of the field: new chapter on cell biology of glycosylation and, expanded coverage of congenital disorders, proteoglycans, influenza virus, muscular dystrophy, and cancer
- Marginal cross-references between the chemical first part and biological and biomedical second part of the book to emphasize the relevance of structure to function
- New and updated special topic boxes represent current research and applications of glycobiology to disease and therapeutics
- Broadened treatment of analytical methods, including glycoarrays.

Acknowledgements

It is again our pleasure to thank our colleagues who have educated us in many aspects of carbohydrate chemistry and biology. We are grateful to many who have provided comments and suggested corrections to previous editions, including Ed Lee at Johns Hopkins University, Michel Monsigny at the University of Orléans, Harry Schachter at the University of Toronto, Nathan Sharon at the Weizmann Institute, Paul Crocker at the University of Dundee, Bill Weis at Stanford University and Madeleine Gentle, Russell Wallis, Daniel Mitchell, and Andrew Fadden at the University of Oxford. We particularly would like to pay tribute to David Wing, who provided many hours of tuition in all aspects of glycobiology, and read and commented on several chapters of the first edition. We also thank Paul Crocker as well as David Bundle at the University of Alberta, David Harvey, Alison Critchley, Tony Merry, Pauline Rudd, and Mark Wormald at the University of Oxford, Andrei Petrescu at the Institute of Biochemistry of the Romanian Academy, and Simon North and Anne Dell at Imperial College London for providing material used to make some of the figures. Finally, we thank Dewi Jackson and the staff of Oxford University Press for suggestions on the format of the third edition and for efficient production of the book.

Contents

Part 1
Structures and biosynthesis of glycans

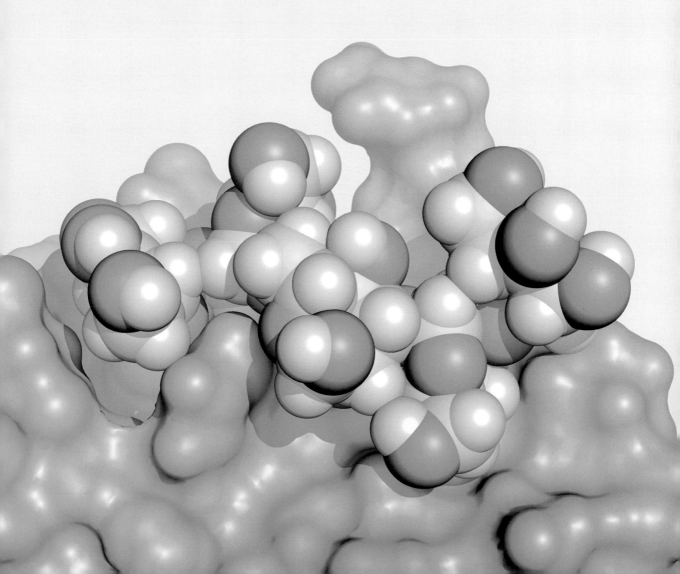

Concepts of glycobiology

LEARNING OBJECTIVES

By the end of this chapter, students should understand:

1   The classes of glycoconjugates and their locations in eukaryotic cells

2   Structures of the sugar monomers found in glycoproteins and glycolipids

3   The nature of glycosidic linkages and the way they are formed

4   The types of function served by glycosylation

The challenge of glycobiology is to define the biological functions of sugars attached to proteins and membranes, and to determine how these functions are carried out. Although this challenge is ongoing, the principles that describe how protein- and lipid-linked sugars mediate biological processes are becoming clear. The goal of this text is to illustrate the themes that underlie the functions of glycoproteins and glycolipids by highlighting well-understood examples of how the carbohydrate portions of these molecules work.

1.1   The field of glycobiology encompasses the multiple functions of sugars attached to proteins and lipids

Most biochemists initially encounter carbohydrates in the context of energy metabolism in cells. The pathways through which energy is abstracted from the breakdown of glucose and glycogen are familiar from introductory biochemistry textbooks. Glycogen is even linked to a protein core and could thus be considered a glycoprotein. However, the function of sugar molecules as storage and transport forms of energy is generally considered to fall outside the field of glycobiology. This distinction is arbitrary but useful, because it allows us to focus on other functions of sugars, many of which are less well understood.

Glycoconjugates are formed when mono-, oligo-, or polysaccharides are attached to proteins or lipids. The sugar-containing portions of the resulting glycoproteins and glycolipids are generally complex heteropolymers rather than repeating homopolymers such as the storage polysaccharides glycogen and amylose. It is common to refer to the sugars of glycoproteins and glycolipids as glycans. Glycans can be built from some of the same building blocks that serve as energy stores, such as glucose, but they also include other monosaccharide units.

There is no single predominant function of protein- and lipid-linked glycans. In this respect, they are similar to proteins themselves, which serve diverse functions as enzymes, hormones, transporters, and structural elements. The functions of glycans fall into at least five broad categories that are summarized in this chapter and discussed in detail in subsequent parts of this book (Figure 1.1). The structural roles of glycoconjugates reflect the physical properties of glycans themselves, so the proteins and lipids to which structural glycans are attached can be viewed as scaffolds that serve an organizational role. Glycans can also affect the intrinsic properties of proteins to which they are attached. In order to function in trafficking, adhesion, and signalling, glycans usually must interact with protein receptors that are known as lectins. The field of glycobiology encompasses the study of glycoconjugates, the enzymes that catalyse their biosynthesis, and the lectins that recognize them.

1.2   There are three major classes of glycoconjugates

The majority of glycans to be considered in this book fall into three groups: those attached to lipids, and those attached to proteins either through a nitrogen atom (N-linked oligosaccharide/glycan) or through an oxygen atom (O-linked oligosaccharide/glycan). Both glycoproteins and glycolipids are found at the extracellular surface of the plasma membrane. In addition, glycoproteins are secreted into biological fluids, such as serum, and they also make up the insoluble extracellular matrix

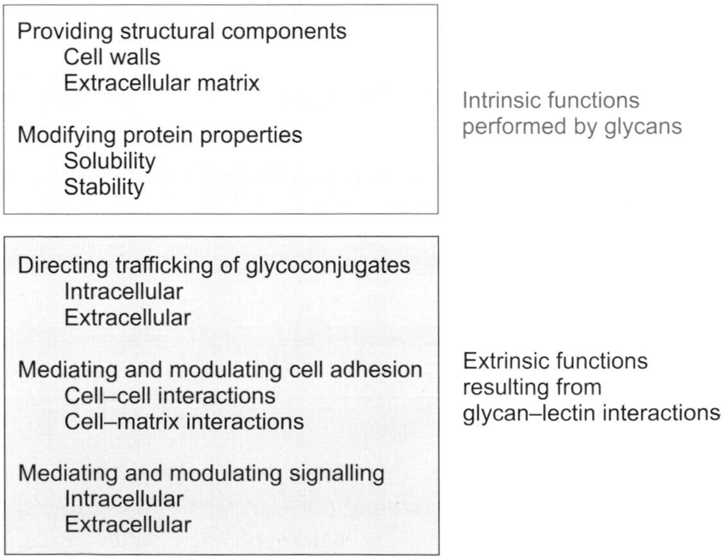

Providing structural components
    Cell walls
    Extracellular matrix

Modifying protein properties
    Solubility
    Stability

Intrinsic functions performed by glycans

Directing trafficking of glycoconjugates
    Intracellular
    Extracellular

Mediating and modulating cell adhesion
    Cell–cell interactions
    Cell–matrix interactions

Mediating and modulating signalling
    Intracellular
    Extracellular

Extrinsic functions resulting from glycan–lectin interactions

Figure 1.1 Summary of some of the functions of glycans.

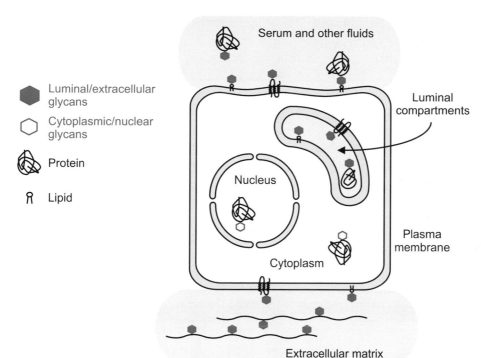

Figure 1.2 Localization of glycoconjugates in intracellular and extracellular compartments.

that surrounds cells (Figure 1.2). All of these groups of molecules have been intensively studied, but the N-linked glycans attached to soluble, secreted proteins are understood best. The emphasis on glycoproteins bearing N-linked oligosaccharides reflects the historical availability of serum glycoproteins as targets for investigation. This group of glycoproteins is used as a point of reference throughout this book. Before they arrive at the cell surface, glycoconjugates must be made inside the cell. This process occurs in the lumen of the endoplasmic reticulum and the Golgi apparatus. Indeed, the steps of glycosylation form an integral part of the secretory machinery of the cell. Thus, most glycoconjugates are separated from the cytoplasm by a membrane. However, partitioning of glycoconjugates into the extracytoplasmic compartments of the cell is not absolute because there are also cytoplasmic and nuclear forms of glycosylation.

1.3   Glycans are composed of monosaccharides with related chemical structures

The most common constituents of glycans are hexoses. Four of the six carbon atoms in a hexose are chiral centres, because each of carbon atoms 2, 3, 4, and 5 is bonded to four chemically distinct structures (Figure 1.3). The substituents around each of these carbon atoms can be arranged in two stereochemically different ways.

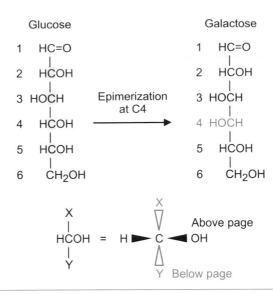

Figure 1.3 Stereochemistry of glucose and galactose. The four asymmetric carbon atoms in glucose are shaded and the convention for showing the stereochemistry is indicated at the bottom.

Because there are four chiral centres, each of which can exist in two configurations, there are a total of 16 possible hexoses. A series of eight common names with the prefixes D- or L- are used to denote these 16 hexoses. For example, D-glucose and D-galactose differ in the configuration of carbon 4 only. A change in the stereo-chemical configuration of a single carbon atom in a sugar is referred to as an epimerization, so D-glucose and D-galactose are epimers. In contrast, D-glucose and L-glucose are mirror images of each other, because the configuration of each of the asymmetric carbons is reversed in these two hexoses. Such pairs of mirror images are called enantiomers.

The hexoses in glycoconjugates are normally found in a six-member ring form known as the pyranose configuration (Figure 1.4). The ring is created by reaction of the 5-hydroxyl group with the 1-aldehyde group to create a hemiacetal. The pyran-ose configuration can be explicitly indicated in the name and abbreviation for a hex-ose, as in D-glucopyranose and D-Glcp. However, it is common to make this designation implicit. In the ring form, the C1 atom is now linked to four chemically distinct atoms, so it is a chiral centre and can exist in either of two stereochemical configurations. C1 is referred to as the anomeric carbon and the two anomeric con-figurations or anomers are distinguished by the designations α and β. C6 is located outside the ring and is thus exocyclic.

Only a few of the possible hexoses are commonly found in glycoconjugates (Figure 1.5). Glucose (Glc) is a convenient reference compound when comparing these structures, as many of them can be derived from glucose by a single epimeriza-tion or substitution. For example, epimerization at the 2 position yields mannose (Man), while epimerization at the 4 position yields galactose (Gal). Substitution of

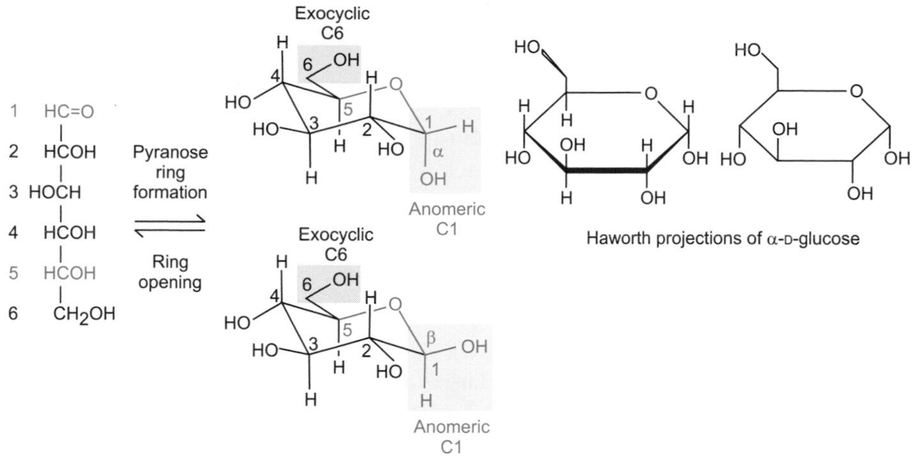

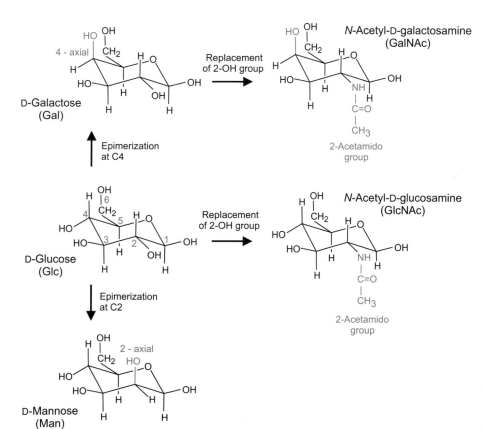

$$
\begin{array}{ll}
1 & HC{=}O \\
2 & HCOH \\
3 & HOCH \\
4 & HCOH \\
5 & HCOH \\
6 & CH_2OH
\end{array}
$$

Pyranose
ring
formation

Ring
opening

Exocyclic
C6

Anomeric
C1

Exocyclic
C6

Anomeric
C1

Haworth projections of α-D-glucose

Figure 1.4 Two ways to form a pyranose (six-member) ring from glucose, in which the anomeric C1 is in either α or β configuration. Simplified Haworth representations of the cyclic forms are sometimes used.

D-Galactose
(Gal)

4 - axial

Epimerization
at C4

Replacement
of 2-OH group

N-Acetyl-D-galactosamine
(GalNAc)

2-Acetamido
group

D-Glucose
(Glc)

Replacement
of 2-OH group

N-Acetyl-D-glucosamine
(GlcNAc)

2-Acetamido
group

Epimerization
at C2

D-Mannose
(Man)

2 - axial

Figure 1.5 Relationships between the common hexoses and *N*-acetylhexosamines.

the 2-hydroxyl group of glucose or galactose with an acetylated amino group yields *N*-acetylglucosamine (GlcNAc) or *N*-acetylgalactosamine (GalNAc). All these hexoses are normally found in the D configuration, so it is common to make this implicit when describing glycan structures.

Additional modified forms of the simple hexoses appear in glycoconjugates, often resulting from changes in C6 (Figure 1.6). For example, oxidation of C6 to a carboxyl group creates a sugar acid such as glucuronic acid (GlcA). Loss of C6 from glucose leads to the generation of xylose (Xyl), which is a pentose. The form of fucose (Fuc) commonly found in mammalian glycoproteins is an interesting case. It is related to galactose by loss of the 6-hydroxyl group, but it is in the L rather than the D configuration. Thus, it is the mirror image of 6-deoxy-D-galactose.

The term sialic acid encompasses a large family of sugars. One member of this family, *N*-acetylneuraminic acid (NeuAc), is the form that is most commonly found in mammalian glycoconjugates (Figure 1.7). NeuAc is a nine-carbon sugar acid formed by condensation of pyruvate with *N*-acetylmannosamine. It is usually found in a six-member ring configuration that is formed by joining the carbonyl group at C2 and the 6-hydroxyl group in a hemiketal. There are several unique substituents extending from the ring, including the carboxyl group at position 1, the acetylated

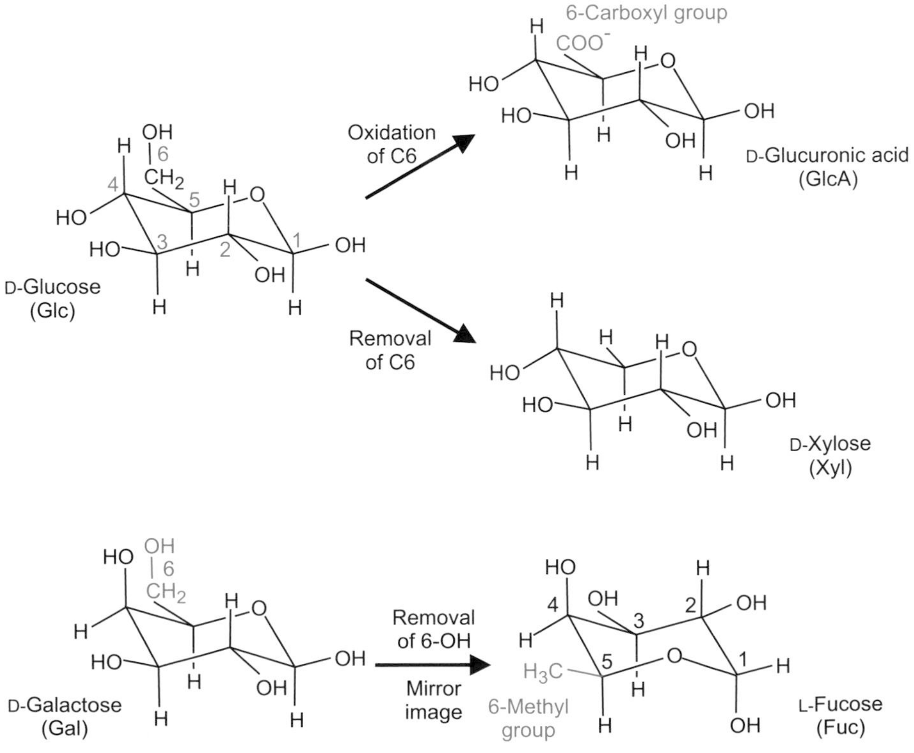

Figure 1.6 Structures of some common derivatives of the hexoses.

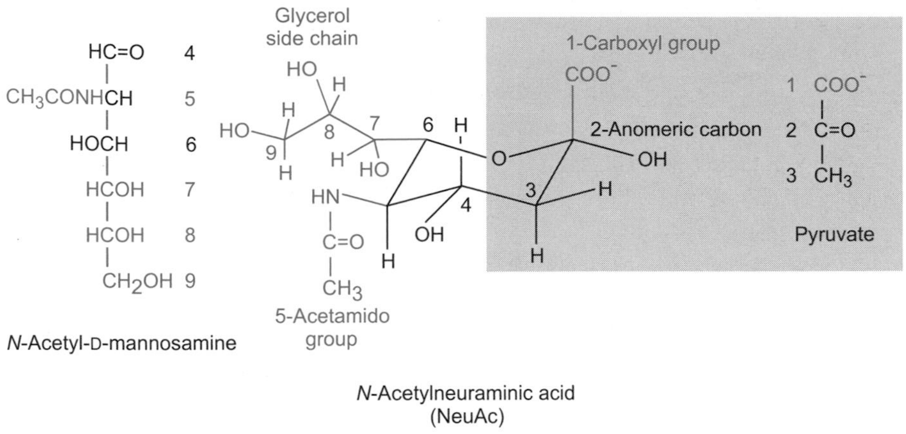

N-Acetyl-D-mannosamine

N-Acetylneuraminic acid
(NeuAc)

Figure 1.7 Structure of N-acetylneuraminic acid, which is the most common form of sialic acid. C1–3 are derived from pyruvate and C4–9 are derived from N-acetylmannosamine.

amino group attached to C5, and the three-carbon chain consisting of C7, C8, and C9. Because each of C7, C8, and C9 is hydroxylated, this substituent is often referred to as the glycerol side chain.

1.4   Glycosidic linkages between monosaccharides exist in multiple configurations

Reaction of a monosaccharide at the hemiacetal of the ring conformation with the hydroxyl group of another monosaccharide results in condensation to form an acetal, with concomitant elimination of a water molecule (Figure 1.8). The resulting structure is a **glycosidic linkage**. As in the free sugars, the anomeric carbon, C1, is linked to four chemically distinct atoms and can exist in α or β configurations.

The structure of a disaccharide can be unambiguously indicated by giving the structures of the constituent monosaccharides and the nature of the linkages. For example, lactose is Galβ1–4Glc. A glycosidic linkage has been formed between the C1 of galactose and the 4-hydroxyl group of glucose, and the anomeric C1 of galactose is in the β configuration. The glucose residue retains the aldehyde function in the form of a hemiacetal. Because of the ability of this group to reduce inorganic ions such as Cu^{2+}, it is referred to as the **reducing end** of the disaccharide. The galactose residue constitutes the **non-reducing end**.

The complete chemical structure of a typical N-linked glycan that might be found in a glycoprotein can be quite cumbersome (Figure 1.9). A description of the monosaccharides and linkages suffices to convey the same information. In spite of the simplifications provided by this nomenclature, it is occasionally useful to provide a still simpler pictorial representation of the structures. By assigning shapes to the various monosaccharides, much of the structural information can be

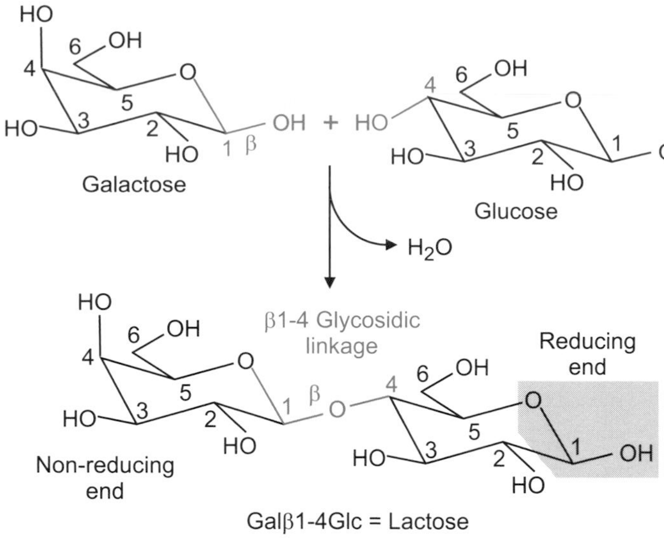

Figure 1.8 Formation of a glycosidic linkage. The glycosidic bond is formed between the reducing end of one monosaccharide (galactose in this case) and one of the other hydroxyl groups of a second monosaccharide (such as the 4-OH group of glucose shown here) by abstraction of a water molecule. After formation of the bond, the only reducing group remaining is at the 1 position of the glucose, which is defined as the reducing end of the disaccharide.

presented in condensed form. Although some information, such as the nature of the linkages, is lost in this representation, this symbol convention is particularly convenient when a structure has been described in detail and is being repeated in modified forms.

1.5   Formation of glycosidic linkages requires energy and is catalysed by specific enzymes

Creation of a disaccharide such as lactose by formation of a glycosidic linkage between the two monosaccharides, galactose and glucose, is an energetically unfavourable process (Figure 1.10). As with many other biochemical processes, the free energy needed to form the bond is generated by coupling the reaction to the energetically favourable hydrolysis of phosphate anhydride bonds. This energy coupling takes place in two stages. The energy of hydrolysis of two high energy phosphate bonds in adenosine triphosphate (ATP) is first used to drive formation of a nucleotide sugar donor, uridine diphosphate (UDP)-galactose (Figure 1.11). UDP-galactose is then used to make the glycosidic linkage with glucose.

Other nucleotide sugar donors can be generated in a similar manner. A glycosyltransferase catalyses the transfer of sugar from the donor and is thus responsible for the formation of the glycosidic linkage. A glycosyltransferase has specificity for a

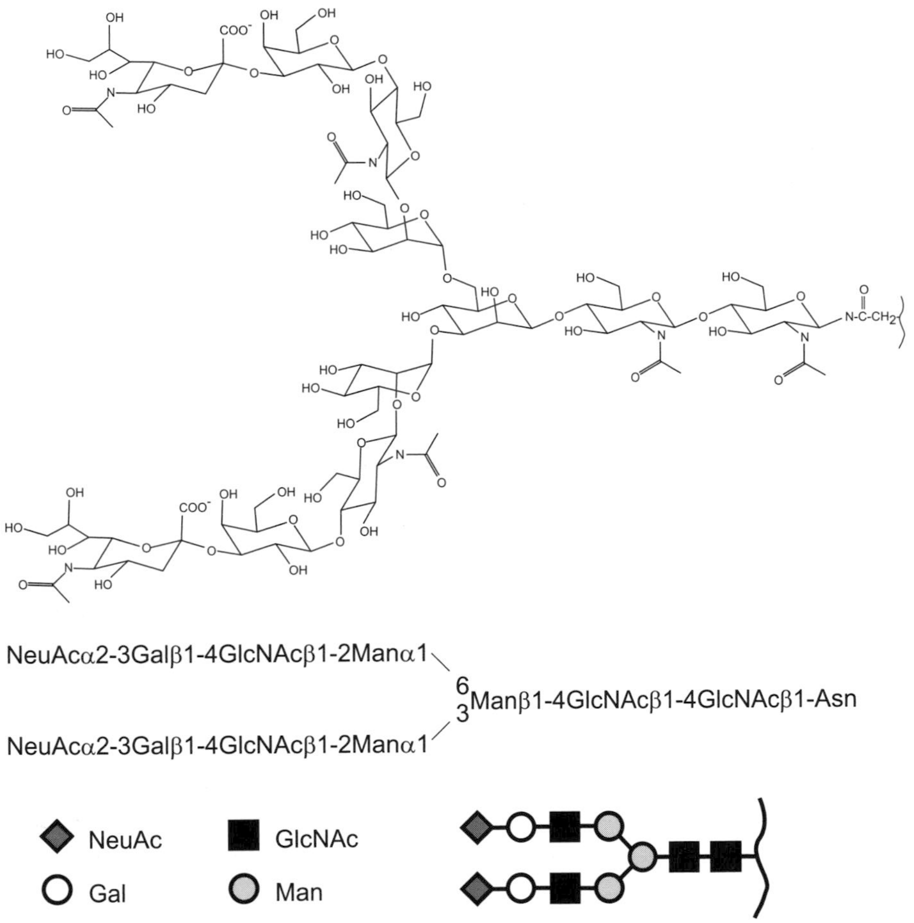

NeuAcα2-3Galβ1-4GlcNAcβ1-2Manα1

$^6_3$Manβ1-4GlcNAcβ1-4GlcNAcβ1-Asn

NeuAcα2-3Galβ1-4GlcNAcβ1-2Manα1

◆ NeuAc   ■ GlcNAc

○ Gal   ● Man

Figure 1.9 A typical N-linked glycan from a glycoprotein presented in a complete chemical structure, word structure, and symbol representation. Although many older publications employ a variety of different symbol codes, common symbols have now been agreed by many investigators and these are employed here. The symbols are particularly useful when full information about the details of linkages is understood implicitly.

nucleotide sugar donor and an acceptor. In the synthesis of lactose, glucose serves as the acceptor molecule. The full name of a glycosyltransferase indicates the nature of the nucleotide sugar donor, the nature of the acceptor, and the nature of the bond formed. Thus, formation of lactose would be catalysed by a UDP-galactose:glucose β1-4-galactosyltransferase. For convenience, a shorter name such as β1-4-galactosyltransferase is usually used.

Because of the strict donor and acceptor specificity of each glycosyltransferase, each enzyme can only add one type of sugar in a specific linkage. For example, different transferases are required for synthesis of lactose and N-acetyllactosamine (Galβ1–4GlcNAc). This concept is often referred to as the 'one enzyme–one linkage' rule. A glycosyltransferase often recognizes not only the specific sugar residue to which it will transfer the donor sugar, but also other features of the oligosaccharide of which the acceptor residue is part. For example, in the biosynthesis of the oligosaccharide shown in Figure 1.9, addition of GlcNAc residues to the two different branches is catalysed by two different GlcNAc-transferases that each recognize more

LIVERPOOL JOHN MOORES UNIVERSITY
LEARNING SERVICES

Overall energetics of glycosidic bond formation

Gal + Glc	$\rightarrow$	Galβ1-4Glc	ΔG = +3.4 kcal/mole
2ATP	$\rightarrow$	2ADP + 2Pi	ΔG = -14.6 kcal/mole

Gal + Glc + 2ATP $\rightarrow$ Galβ1-4Glc + 2ADP + 2P$_i$ ΔG = -11.2 kcal/mole

Synthesis of nucleotide sugar donor

ATP + Gal	$\rightarrow$	ADP + Gal-1P
ATP + UDP	$\rightarrow$	ADP + UTP
UDP-Glc + Gal-1P	$\rightarrow$	UDP-Gal + Glc-1P
Glc-1P + UTP	$\rightarrow$	UDP-Glc + PP$_i$
PP$_i$	$\rightarrow$	2P$_i$

Gal + UDP + 2ATP $\rightarrow$ UDP-Gal + 2ADP + 2P$_i$

Creation of glycosidic bond

UDP-Gal + Glc $\rightarrow$ Galβ1-4Glc + UDP

Figure 1.10 Energetics of formation of a glycosidic bond. Energetically unfavourable creation of the glycosidic bond is ultimately driven by the hydrolysis of two phosphate linkages in ATP. The reactions are linked by synthesis of UDP-galactose. Energy originally derived from ATP is released when the nucleotide sugar serves as the sugar donor during formation of the glycosidic bond. ADP, adenosine diphosphate; ATP, adenosine triphosphate; PP$_i$, pyrophosphate; UDP, uridine diphosphate; UTP, uridine triphosphate.

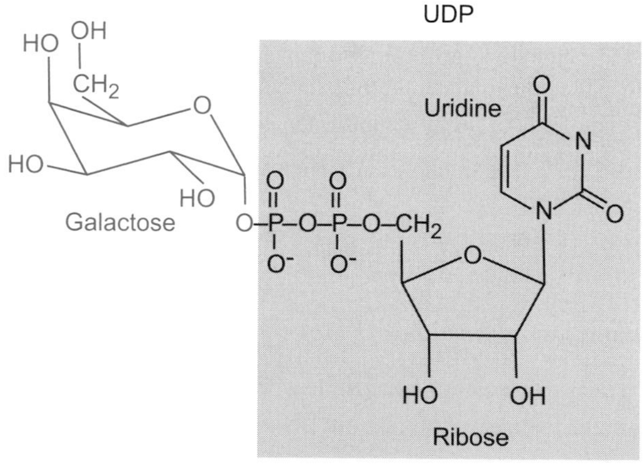

Figure 1.11 Structure of a nucleotide sugar that can serve as a sugar donor in a glycosyltransferase reaction. UDP, uridine diphosphate.

than just the mannose acceptor residue. In contrast, the two galactose and sialic acid residues can be added by a single type of galactosyltransferase and sialyltransferase. Although each glycosyltransferase is highly specific, there are often several different glycosyltransferases with similar or overlapping specificities. For example, in humans there are six sialyltransferases that can each add sialic acid in 2–3 linkage to galactose.

Breaking of glycosidic linkages also requires specific enzymes, but does not require input of energy. Glycosidases catalyse the energetically favourable hydrolysis of glycosidic linkages. Like glycosyltransferases, glycosidases are specific, with each enzyme only catalysing hydrolysis of glycosidic linkages involving a particular sugar. For example, a sialidase (neuraminidase) can catalyse release of a sialic acid residue (NeuAc) from the non-reducing terminus of an oligosaccharide such as the one shown in Figure 1.9. Glycosidases can also be specific for particular linkages of a sugar. Taking sialidase as the example again, some sialidases will only hydrolyse α2–3 linkages of NeuAc, while others do not discriminate between α2–3 and α2–6 linked NeuAc.

1.6   Understanding structure–function relationships for glycans can be more difficult than for other classes of biopolymers

Comparison with proteins serves as a good starting point for thinking about how glycans carry out useful biological functions. In spite of their diverse biological roles, proteins share two common features that unify the study of their properties: each protein is synthesized as an identical copy by translation of a messenger RNA template that is encoded in the genome and the activity of a protein results from formation of a precisely folded three-dimensional structure. In contrast, glycans are assembled without a template through a series of individually catalysed reactions. The resulting structures are not unique, because many different proteins are modified with a common set of glycan structures and different copies of a single polypeptide backbone can be modified with scores of distinct glycans. In addition, glycans often appear to lack a discrete, folded structure. All of these features make it difficult to establish structure–function relationships for glycans and some novel principles must be defined to describe the ways that glycans function.

One explanation for the lack of simple rules about which specific glycans are attached to specific proteins is that the functions of the protein and glycan portions of many glycoproteins can be independent of each other. That is, all copies of a particular protein perform the same function regardless of what glycans are attached and all copies of a particular glycan perform the same function, although they are attached to different proteins (Figure 1.12). This arrangement can be particularly effective when glycans serve as tags that can be recognized and used to direct glycoprotein trafficking. For example, glycoproteins in the secretory pathway of eukaryotic cells are subjected to a series of quality control checks. The common glycans

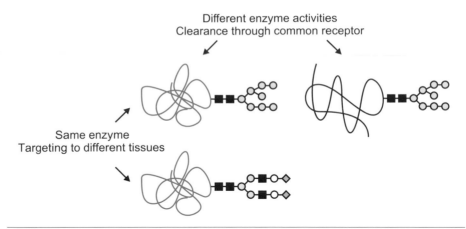

Figure 1.12 Potential independent functions of polypeptide and glycan portions of glycoproteins. Different proteins can bear identical glycans and different copies of one protein can be heterogeneously glycosylated with multiple glycans. Thus, targeting to different sites directed by the glycans may be independent of the enzymatic or other functions of the proteins.

attached to the majority of these proteins are used to hold them in the appropriate luminal compartments during this process. One set of glycans can serve this function for a multitude of different secretory glycoproteins in spite of the fact that the proteins themselves will have diverse functions once outside the cell. In such cases, the glycan portion of the mature glycoprotein may have essentially no role once the protein has reached the cell surface.

In other cases, the independent functions of protein core and glycan decoration may be manifest in a different way. When a particular glycan is attached to a protein or lipid at the extracellular surface of the plasma membrane, it may mediate adhesion or anti-adhesion events independently of the carrier lipid or protein to which it is attached. The presence of similar glycan structures on a variety of membrane glycoproteins and glycolipids may provide a mechanism for achieving high densities of these structures without requiring a correspondingly high density of any one type of membrane protein or lipid. Thus, in some cases, glycans and the proteins or lipids to which they are attached can be studied more or less in isolation from each other. However, in other cases the role of a particular glycan is only evident in the context of a specific glycoconjugate.

1.7   Glycan structures are encoded indirectly in the genome

Genomic DNA sequences dictate the structures of glycoconjugates just as they determine the structure of all cell components. The sugar structures are not encoded directly in the DNA sequences, but are determined by transcription and translation of genes to generate glycosyltransferases that in turn control synthesis of the glycan portions of glycoconjugates (Figure 1.13). Thus, compared with the biosynthesis of proteins, there is an extra step in the decoding process. The one enzyme–one linkage

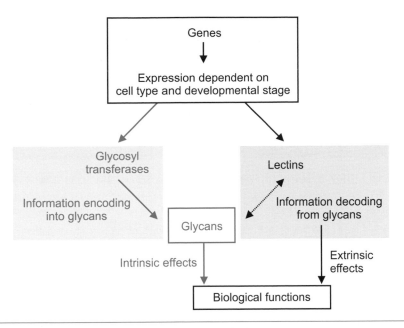

Figure 1.13 Steps in encoding and decoding information in glycan structures.

rule suggests that it will eventually be possible to describe the full repertoire of glycan structures that can be made in a particular cell by determining which glycosyltransferases are expressed in this cell.

An important tool in understanding both the synthesis of glycoconjugates and their functions has been the generation of knockout mice in which glycosyltransferases have been eliminated. Results of many specific experiments of this type are cited throughout this book. However, two important general results are worth noting at this point. First, complete elimination of any of the classes of glycoconjugates is fatal to the organism at an early stage in development, demonstrating that these molecules perform critical biological functions for the organism. Second, in almost all cases, cells lacking any of these classes of glycoconjugates are viable. Thus, although a number of important roles of glycoconjugates are discussed in a cellular context, many of their functions are organismal rather than cellular.

SUMMARY

Because glycan structures are created by glycosyltransferases and are recognized by lectins, glycobiology encompasses the study of these proteins as well as the glycans themselves. There are a large number of possible hexoses and related structures, but only a small fraction of these monosaccharide units are found in glycoconjugates. Similarly, the monosaccharide units could potentially be combined in very many different ways through different linkages, but the glycosyltransferases catalyse synthesis of only a limited number of the possible structures. Nevertheless, there is great diversity in the glycan portions of glycoconjugates. It is often difficult to assign functions to specific glycans because the functions may only be evident in an organismal context and gene knockout approaches may be required to probe these roles.

KEY REFERENCES

Alper, J. (2001). Searching for medicine's sweet spot, *Science* **291**, 2338–2343. This issue of *Science* highlights the field of glycobiology with several reviews and features about recent advances.

Drickamer, K. and Taylor, M.E. (1998). Evolving views of protein glycosylation, *Trends in Biochemical Sciences* **23**, 321–324. Evolution is used to relate the different functions of glycosylation to each other.

Lis, H. and Sharon, N. (1993). Protein glycosylation: structural and functional aspects, *European Journal of Biochemistry* **218**, 1–27. In addition to summarizing what glycans are attached to proteins, these authors provided early speculation on some of their functions.

Roseman, S. (2001). Reflections on glycobiology, *Journal of Biological Chemistry* **276**, 41527–41542. Reflecting on the development of carbohydrate biochemistry, a veteran of the field provides perspective on how we now think about the functions of glycosylation.

Varki, A., Cummings, R.D., Esko, J.D., Freeze, H.H., Stanley, P., Bertozzi, C.R., Hart, G.W., and Etzler, M.E. (2008). *Essentials of Glycobiology* (2nd edn). Cold Spring Harbor, NY: Cold Spring Harbor Laboratory Press. This comprehensive book on glycobiology provides extensive detail on many topics along with a wealth of references.

QUESTIONS

1.1   Compare the structural features of proteins and oligosaccharides.

1.2   What are the main types of functions performed by glycans attached to glycoproteins and glycolipids? Discuss why glycans might be better suited than proteins to performing some of these functions.

N-Linked glycosylation

LEARNING OBJECTIVES

By the end of this chapter, students should understand:

1 The structures of different classes of N-linked glycans

2 Chemical steps in the biosynthesis of N-linked glycans

3 The functions of glycosyltransferases

4 The cellular compartmentalization of biosynthetic steps

5 The nature and causes of heterogeneity of N-linked glycans

The N-linked glycosylation pathway is the best understood route to protein glycosylation. This pathway can be used to illustrate general principles of glycoconjugate structure and biosynthesis. The nature of the N-linked glycosylation pathway also provides some insight into the functions and evolutionary history of glycosylation.

2.1 Diverse N-linked glycans have a common core structure

N-linked glycans in glycoproteins are attached to the amide nitrogens of asparagine side chains. In animal cells, the sugar linked to an asparagine residue is almost inevitably *N*-acetylglucosamine (GlcNAc) and the linkage is always in the β configuration (Figure 2.1). The arrangement is similar to a glycosidic linkage, except that the anomeric carbon is bonded to the amide nitrogen rather than to a sugar hydroxyl group. This linkage can be abbreviated as GlcNAcβ1-Asn. The different categories of N-linked glycans, such as high-mannose oligosaccharide/glycan and complex types, contain a common core structure but differ in the terminal elaborations that extend from this core.

2.2 Assembly of N-linked glycans occurs in three major stages

It is convenient to divide the biochemical pathway for N-linked glycan synthesis into three stages:

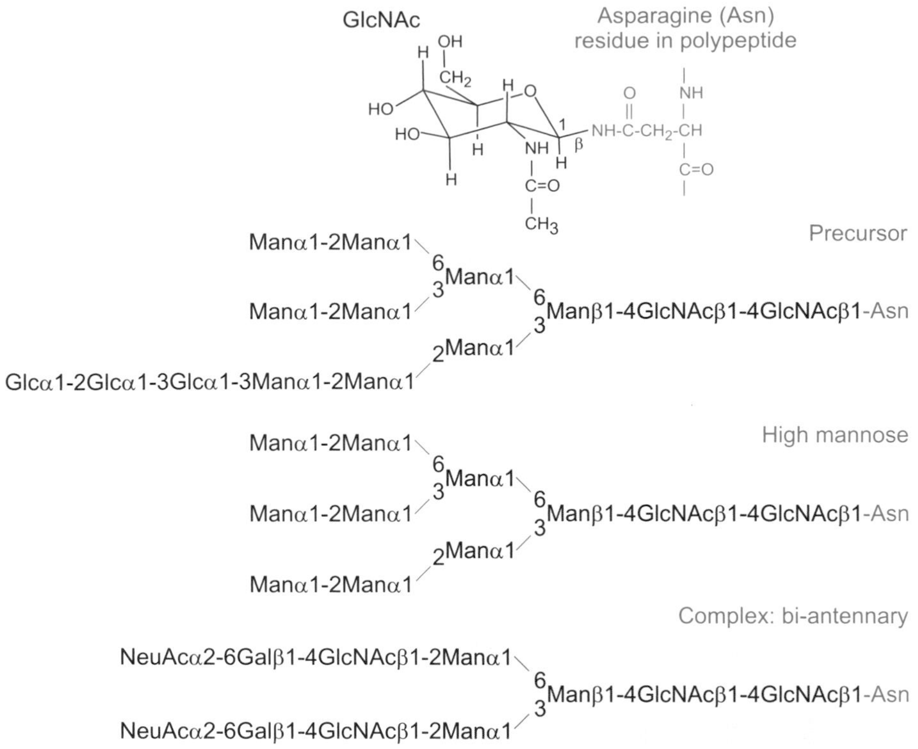

Figure 2.1 Attachment of an N-linked glycan to an asparagine residue and some examples of N-linked glycans. The common core structure is shaded. Because the two branching mannose residues in the core are in different linkages to the innermost mannose residue, these two branches are known as the 1–3 arm and the 1–6 arm of an oligosaccharide respectively.

- formation of a lipid-linked precursor oligosaccharide
- *en bloc* transfer of the oligosaccharide to the polypeptide
- processing of the oligosaccharide.

Processing steps include removal of some of the original sugar residues (trimming), followed by addition of new sugars at the non-reducing termini of the glycan. Within a cell, synthesis of glycoproteins takes place in spatially differentiated steps (Figure 2.2). Lipid-linked precursor synthesis, *en bloc* transfer to the protein, and initial trimming reactions occur in the rough endoplasmic reticulum, while subsequent processing steps occur as the nascent glycoprotein migrates through the Golgi apparatus.

2.3   The precursor oligosaccharide for N-linked glycans is assembled on the lipid dolichol

The donor that initiates N-linked glycan synthesis is a $Glc_3Man_9GlcNAc_2$ structure attached to the lipid **dolichol** through a pyrophosphate linkage (Figure 2.3).

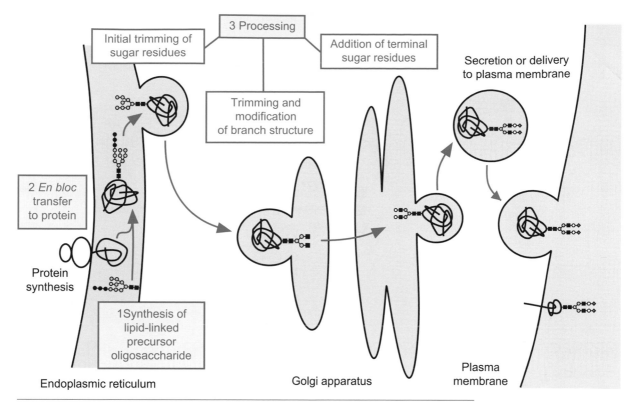

Figure 2.2 Overview of the pathway for glycoprotein biosynthesis and its location within a cell. Early stages of biosynthesis, in which a glycan is assembled on a lipid and transferred to the glycoprotein, occur in the endoplasmic reticulum. The glycan is then processed by glycosidases that remove some of the sugars and glycosyltransferases which add additional residues to form branches and diverse terminal structures. The processing events occur in specific compartments within the endoplasmic reticulum and Golgi apparatus.

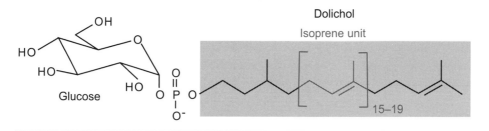

Figure 2.3 A dolichol sugar that can serve as the donor in glycosyltransferase reactions on the luminal side of the endoplasmic reticulum membrane.

The structure of dolichol reflects its synthesis by condensation of five-carbon isoprene units, such as those that are combined to form cholesterol and other sterols. However, in dolichol the units are assembled end-to-end and are not cyclized. The presence of 19 isoprene units means that the hydrophobic portion of this lipid is far

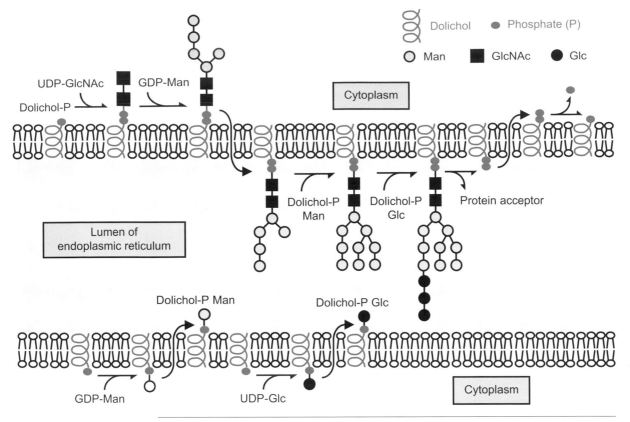

Figure 2.4 Pathway for generation of the dolichol-linked oligosaccharide donor for protein N-glycosylation. Seven sugar residues are assembled on dolichol facing the cytoplasmic side of the membrane, after which the entire structure is flipped to face the luminal compartment. Sugar donors for glycosyltransferases that make further additions on the luminal side of the membrane are also linked to dolichol and are flipped across the membrane. After the glycan is attached to the protein, released dolichol pyrophosphate must flip back to face the cytoplasm to complete the cycle.

longer than the fatty acid tails on membrane phospholipids. Nevertheless, this extended hydrophobic region is inserted into the lipid bilayer, possibly in a helical or folded conformation.

Assembly of a glycan on the dolichol head group takes place in two phases. The first phase occurs on the cytoplasmic side of the endoplasmic reticulum membrane, whereas the second phase occurs in the lumen (Figure 2.4). The enzymes that catalyse attachment of the two GlcNAc residues and the first five mannose residues utilize nucleotide sugar donors UDP-GlcNAc and guanosine diphosphate (GDP)-Man directly. At this point, the lipid-linked glycan is translocated across the membrane and becomes inaccessible to cytoplasmic enzymes. The mechanism of this translocation remains very poorly understood.

Once the growing glycan chain is exposed on the luminal side of the endoplasmic reticulum membrane, further sugars are added. The immediate donors for the addition of the final four mannose residues and the three glucose residues are dolichol-linked sugars. These lipid-linked monosaccharides are synthesized on the cytoplasmic face of the endoplasmic reticulum membrane by reaction of dolichol phosphate with UDP-Glc or GDP-Man. Translocation of the dolichol phosphomonosaccharides across the membrane to face the lumen is analogous to translocation of the oligosaccharide–dolichol intermediate and is again poorly understood. The energy needed for synthesis of the glycosidic linkages in these reactions comes from the saccharide–phosphate bonds through which the monosaccharides are linked to dolichol phosphate.

Transfer of the completed glycan to nascent polypeptides also occurs on the luminal side of the endoplasmic reticulum membrane. The energy required for synthesis of the GlcNAc-asparagine linkage is again provided by the energy derived from cleavage of the glycan–phosphate bond. Because the glycan is attached to dolichol with a pyrophosphate head group, the action of a phosphatase is required to regenerate dolichol phosphate and thus complete the dolichol cycle.

2.4   The dolichol-linked precursor oligosaccharide is transferred to asparagine residues of polypeptides

In general, asparagine residues to which glycans are to be attached must meet three conditions:

- they must be located in a specific sequence context within the primary structure of the protein
- they must be located appropriately in the three-dimensional structure of the protein
- they must be found in the correct intracellular compartment.

Glycosylated asparagine residues are almost invariably found in the sequences Asn-X-Ser or Asn-X-Thr, where X can be any amino acid except proline, although very rare instances in which the asparagine in the sequence Asn-X-Cys or other contexts becomes glycosylated have also been described. N-linked glycans are found at the surfaces of proteins and are not buried within them. Finally, N-linked glycosylation is initiated exclusively in the lumen of the endoplasmic reticulum. Therefore, target asparagine residues are found in secretory proteins and in the portions of transmembrane proteins that face the lumen.

The enzyme that catalyses transfer of the completed dolichol-bound precursor glycan to a polypeptide acceptor is oligosaccharyltransferase (Figure 2.5). The properties of this enzyme readily explain the first and last of the conditions for glycosylation. The requirement for the Asn-X-Ser/Thr acceptor sequence is determined by the specificity of this enzyme. Similarly, the fact that N-linked glycans are found exclusively in luminal portions of proteins reflects the luminal location of this enzyme.

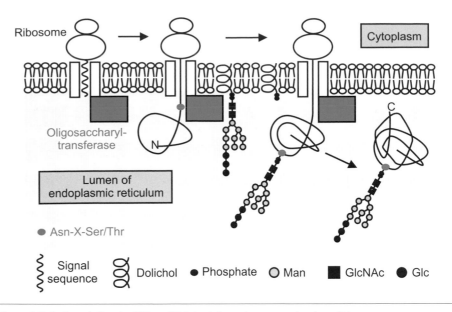

Figure 2.5 Co-translational addition of N-linked glycan to a nascent polypeptide. Oligosaccharyltransferase is associated with the channel through which the polypeptide is translocated to the endoplasmic reticulum, so glycosylation occurs while the polypeptide is still unfolded. However, the glycan is found at the surface of the final, folded protein.

The second condition for glycosylation, that it occurs exclusively at the surface of proteins, makes good energetic sense because the relatively polar nature of the sugars would be incompatible with the hydrophobic interior of proteins. However, the mechanism underlying achievement of this condition is unclear. Although glycosylation is often referred to as a post-translational modification, one form of oligosaccharyltransferase is associated with components of the endoplasmic reticulum membrane that bind ribosomes and translocate the polypeptide through the membrane into the luminal space. Thus, glycan transfer usually takes place as the nascent polypeptide emerges in the lumen of the endoplasmic reticulum and is a co-translational event. A second form of oligosaccharyltransferase glycosylates missed target sites post-translationally. Either way, glycosylation usually takes place on an unfolded polypeptide and one might imagine that any Asn-X-Ser/Thr sequence would be a potential target. While it is true that such relatively polar amino acids are often destined to appear at the surface of proteins, instances are known in which such sequences are buried and are not glycosylated. It is possible that in these cases rapid folding precludes access by oligosaccharyltransferase.

Following transfer to the polypeptide, the N-linked glycan is processed, first by removal of some of the sugar residues by processing glycosidases. These enzymes catalyse the energetically favourable breaking of the glycosidic linkage by addition of a water molecule. These enzymes are exoglycosidases, working only on monosaccharide residues located at the non-reducing termini of the glycan. The first processing steps,

⊙ See Section 1.5 for more on glycosidic linkages.

trimming of the three glucose residues, take place while the initially glycosylated polypeptide is still in the endoplasmic reticulum. Glucosidase I is responsible for removal of the terminal α1-2-linked glucose residue, while glucosidase II removes the two inner α1-3-linked glucose residues once the terminal residue has been removed. Removal of the final glucose residue signals that the newly glycosylated glycoprotein is ready for transit from the endoplasmic reticulum. The quality control system that monitors protein folding is associated with this processing reaction.

⊘ See Sections 10.2 and 10.3 for more on glycoprotein quality control.

2.5   The core oligosaccharide structure is modified by glycosidases and glycosyltransferases

The structure remaining after the action of glucosidases I and II is subject to the action of a series of mannosidases that remove some or all of the four mannose residues in α1-2 linkages (Figure 2.6). Processing is initiated in the endoplasmic reticulum and continues in the *cis* portion of the Golgi apparatus. The glycans attached to

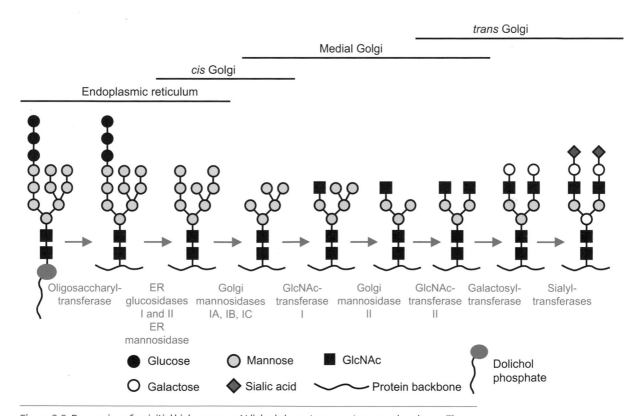

Figure 2.6 Processing of an initial high mannose N-linked glycan to generate a complex glycan. The primary pathway for processing involves an ordered series of steps that occur in different compartments within the cell. Alternative pathways of glycosidase digestion provide ways to bypass some of the steps and additional glycosyltransferases result in addition of further branches and different terminal structures. ER, endoplasmic reticulum.

◐ For more on intracellular localization of glycosyltransferases see section 5.6.

some proteins remain in this state as the glycoproteins move though various luminal compartments to the cell surface. Such glycans, containing between five and nine mannose residues, are called high mannose oligosaccharides. Other glycans are processed to more complicated structures.

Complex glycans are ultimately built on a core that consists of just three mannose residues and two GlcNAc residues attached to the glycoprotein. However, the rebuilding process is initiated by attachment of a GlcNAc residue to the 1–3 arm of the core while it still contains five mannose residues. The enzyme with the capacity to initiate re-elongation, GlcNAc-transferase I, is located in the medial portion of the Golgi apparatus. Following this addition, mannosidases in the Golgi apparatus remove two additional mannose residues from the 1–6 arm of the core. Further additions of GlcNAc, each catalysed by a distinct GlcNAc-transferase, determine the ultimate structure of the glycan by defining a branch structure that is built on to the core. Although bi-antennary glycans are most abundant, tri-antennary and tetra-antennary glycans are also common (Figure 2.7). Rarer glycans containing five or more branches have been documented. The branched structures are typically extended by the addition of a single galactose and sialic acid residue to each GlcNAc residue. The galactosyltransferase and sialyltransferases responsible for these steps are located further along in the secretory pathway in the *trans* portion of the Golgi apparatus and the *trans*-Golgi network. The galactose residue is often in β1-4 linkage to the underlying GlcNAc and the terminal sialic acid can be in either α2-3 or

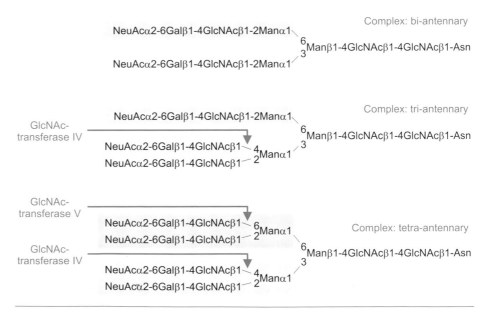

Figure 2.7 Some of the common complex N-linked glycans. Additional forms can exist, including forms with more branches and isomeric forms such as a tri-antennary oligosaccharide in which two branches extend from the 1–6 arm of the core rather than the 1–3 arm. Specific GlcNAc-transferases that initiate formation of different branches are indicated.

α2-6 linkage. The resulting *N*-acetylneuraminic acid-galactose (NeuAc-Gal) disaccharide is often referred to as a capping structure.

The donors for the additions in complex oligosaccharide biosynthesis are all nucleotide sugars (UDP-GlcNAc, UDP-Gal, and cytidine monophosphate-NeuAc (CMP-NeuAc)). This raises an important topological issue because all these reactions occur in the lumen of the Golgi apparatus, yet the donors are made in the cytoplasm. There is no translocation intermediate analogous to dolichol phosphomannose used in dolichol precursor biosynthesis. Instead, transporter molecules facilitate entry of the donors into the luminal compartment. These transporters act as antiporters. For example, following transfer of GlcNAc on to a glycan chain, the UDP by-product is hydrolysed to uridine monophosphate (UMP) and inorganic phosphate. As UDP-GlcNAc is transported into the lumen, UMP is transported back into the cytoplasm.

◉ See **Section 5.5** for more on transporters.

2.6 Hybrid structures and polylactosamine sequences are common extensions of the core oligosaccharide

Many variations are seen in the terminal residues of different N-linked glycans, but there are also some common variants of the core oligosaccharide. For example, complex glycans can be modified by addition of a bisecting GlcNAc residue to the 4 position of the core mannose residue (Figure 2.8). A core fucose residue can also be linked to the GlcNAc residue that is directly attached to the asparagine residue of the protein. Hybrid structures are formed when Golgi processing mannosidases do not act after the addition of a single GlcNAc residue by GlcNAc-transferase I. Processing of the GlcNAc-terminated branch can continue, as in the case of complex glycans, generating a structure containing terminal galactose or sialic acid as well as mannose.

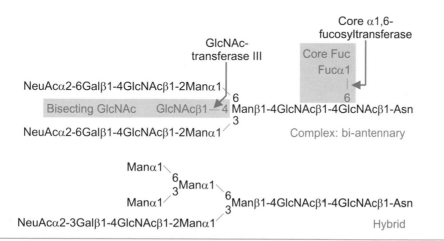

Figure 2.8 Variations on N-linked glycan structures. In addition to diversity resulting from increased branching, some glycans contain additions to the core or are partially processed from high mannose to complex forms.

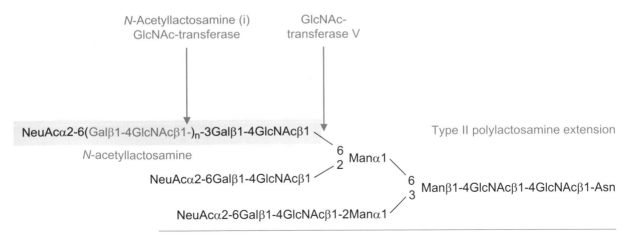

Figure 2.9 One possible form of an N-linked glycan with a polylactosamine extension. Polylactosamine chains are commonly added to a 1–6 branch initiated by the action of GlcNAc-transferase V.

Perhaps the most striking variation on the core structure is the extension of one branch by alternating galactose and GlcNAc residues (Figure 2.9). This sequence can be viewed as a series of repeated Gal-GlcNAc disaccharides and is called a **polylactosamine extension**. The linkages between the sugars can vary, so different families of polylactosamine extensions exist. In the most common form, designated type II, the linkages are Galβ1-4GlcNAc and GlcNAcβ1-3Gal. There is often only a single polylactosamine extension on a given oligosaccharide and it is usually attached to a specific branch. Synthesis of this structure results from the action of a special GlcNAc-transferase that is able to use the galactose-terminated branch as an acceptor. Following addition of the GlcNAc residue, repeated action of the GlcNAc-transferase and a galactosyltransferase generates the polylactosamine sequence. In contrast to the type II chains, the linkage within each lactosamine unit is Galβ1-3GlcNAc in type I chains. Addition of a sialic acid residue to the terminal galactose residue would prevent further elongation of the chain such that the length of the final structure results from the balance between the activities of the GlcNAc-transferase and the sialyltransferase.

2.7   ABO blood groups are determined by the presence of different terminal sugars on glycans of red blood cells

The structures of terminal sugars on different glycans are a continuing theme throughout this book. In this section, a few examples are introduced to illustrate the diverse nature of these often elaborate structures.

Perhaps the best known glycan structures are the ABO blood group substances found on the outer surface of erythrocyte membranes. The majority of the oligosaccharides that form the A antigen and the B antigen are N-linked glycans on band 3,

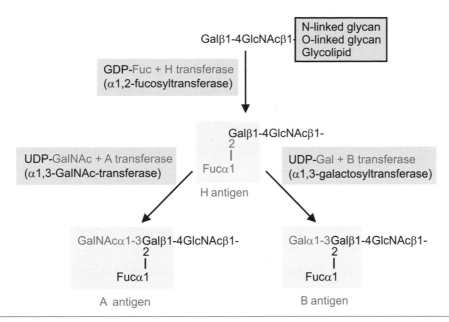

Figure 2.10 Terminal structures associated with the ABO blood group system. Following addition of fucose to the common Galβ1-4GlcNAc disaccharide in glycoproteins and glycolipids, the structures of the final glycans depend on whether an individual expresses either or both of the A or B transferases. The fucose attached in this case is denoted an outer arm fucose to distinguish it from core fucose (Figure 2.8).

the erythrocyte anion transport protein, although similar structures can be found on glycolipids as well. The antigenic glycans are built on the ends of polylactosamine chains by the action of a fucosyltransferase to generate the H substance (Figure 2.10). In individuals of type O blood group, this is the only modification. In A-type individuals, an *N*-acetylgalactosamine (GalNAc) residue is also added to the terminal galactose residue, while in B-type individuals, a Gal residue is appended.

The two different glycosyltransferases responsible for the addition of Gal and GalNAc are encoded by different alleles at a single genetic locus on chromosome 9. In O-type individuals, null alleles are found at this locus in both copies of chromosome 9, so there is no functional transferase and these individuals express only the H substance. The A-type allele encodes a GalNAc-transferase and the B allele encodes a galactosyltransferase. Hence, individuals with two A alleles or one A and one null allele will display GalNAc-terminated glycans, individuals with two B or one B and one null allele will display Gal-terminated glycans and individuals with one A and one B allele will display both types of glycans. The allelic GalNAc- and Gal-transferases are extremely similar to each other: there is only one essential amino acid difference.

All individuals are exposed to the A, B, and H structures from an early age (mostly in food substances), but the immune response that results is modulated by the nature of the structures expressed on each individual's own cells. A-type individuals develop antibodies against the B structure but are tolerant to the A structure, while the reverse

Glycobiology and disease
Blood groups and transfusion

is true for B-type individuals. O-type individuals make antibodies to both A and B structures, while AB individuals make antibodies to neither. Transfused cells must not express glycans to which the recipient has antibodies. For example, an A-type individual can accept cells from an A-type or an O-type donor, but not from B-type or AB-typedonors, because the latter two cell types express B-type terminal structures. The H substance was originally defined antigenically because certain rare individuals lack the fucosyltransferase and hence do not make the H, A, or B structures. These individuals produce antibodies that react with the H structure and are referred to as having the Bombay phenotype.

2.8   The N-linked glycans of an individual glycoprotein are usually heterogeneous

◉ See Section 6.5 for more on glycan numbers.

An important aspect of the lack of correspondence between glycan and protein structures is that an asparagine residue in different copies of a glycoprotein can be modified with different glycans. Glycoprotein molecules with a common polypeptide chain but bearing different glycans are called glycoforms. Because there might be a thousand or more possible glycans, there could be several thousand glycoforms of a glycoprotein that has multiple N-glycosylation sites. In fact, there are typically fewer than a hundred glycoforms of most glycoproteins and usually only a few of these predominate. However, the extent of this heterogeneity varies from protein to protein.

At one extreme, certain sites are modified with one unique type of oligosaccharide. For example, all copies of the soya bean agglutinin protein have the $Man_9GlcNAc_2$ oligosaccharide that has not been further processed. In many proteins, particular asparagine residues are occupied by particular classes of glycans. The single site of glycosylation in chicken ovalbumin is always of the hybrid type or high mannose type, and pancreatic ribonuclease B contains a single site of glycosylation that is modified with high mannose oligosaccharides which range from three to nine mannose residues. However, ribonuclease exhibits another form of heterogeneity because it can also be secreted in an unglycosylated form, known as ribonuclease A. A large number of glycoproteins bear complex N-linked glycans that exhibit heterogeneity in the degree of branching, the presence or absence of core fucose and bisecting GlcNAc residues, and the presence or absence of polylactosamine extensions. In different copies of some glycoproteins, a particular asparagine residue can be derivatized with complex, hybrid, or high mannose glycans.

There are several different ways in which to view the significance of the heterogeneity of glycosylation. It seems most likely that, in many cases, heterogeneity exists because it does not matter. The glycan either serves no specific function, or it is able to serve its function through the common or conserved part of the glycan. Like the many amino acid polymorphisms in proteins, the variability in glycosylation (although more extreme) would reflect essentially a neutral process of evolution. Alternatively, the heterogeneity might be subject to positive selection, at least in some instances. Glycosylation can affect the properties of some proteins and differential glycosylation could affect these properties differentially. However, it is difficult to demonstrate that the range of behaviour generated by heterogeneous glycosylation serves a biologically important role. Finally, it has been suggested that evolutionary pressure for heterogeneous glycosylation of cell surfaces might be generated by the fact that viruses and toxins often use glycans as a means of attaching to cells. Although a more diverse set of cell surface glycans might provide a less uniform target for pathogens, the phenomenon is again difficult to quantify.

➲ See **Chapter 8** for more on the effects of glycosylation on proteins.

2.9 The nature of N-linked glycans attached to an individual glycoprotein is determined by the protein and the cell in which it is expressed

Diversity of glycosylation is controlled by several factors. First, although there is no simple one-to-one correlation between a polypeptide sequence and the structures of glycans attached to it, the structure of a protein can have some influence on its glycosylation. Common N-linked structures, including high mannose and complex oligosaccharides, are found on many glycoproteins and appear to represent a default form of glycosylation, but certain terminal elaborations are found only on subsets of glycoproteins. Modification of the glycans attached to these proteins is specified, at least in part, by structural features of the protein. Examples discussed in later chapters include the attachment of mannose 6-phosphate, GalNAc 4-SO$_4$, and blood group substances to glycans. Secondly, the position of certain glycosylation sites within some proteins favours processing of N-linked glycans to complex oligosaccharides, whereas glycans attached at other sites remain as high mannose structures. Finally, the same protein expressed in different cell types, or in a single cell type under different growth conditions, can be glycosylated in different ways. Some differences may arise from different sets of glycosyltransferases that are expressed in different cell types. Other differences may reflect the different rates at which glycoproteins transit through the secretory pathway, which would determine how long they encounter various parts of the glycosylation machinery. Extreme examples of differential glycosylation are encountered when normal tissue cells are compared with tumour cells derived from these same tissues. Cells from different species will also attach different glycans to similar proteins on their surfaces.

➲ See **sections 10.5 and 10.9** for biological roles of mannose 6-phosphate and GalNAc 4-SO$_4$.

➲ See **sections 13.7 and 13.8** for more on tumour cell glycosylation.

2.10 High mannose structures are present in lower eukaryotes, but the glycosylation machinery has evolved to produce complex glycans in higher organisms

A striking feature of the pathway for N-linked glycosylation is that it does not seem to be a very efficient way to generate many of the glycans attached to glycoproteins.

The route for synthesis of the complex glycans seems particularly circuitous. Two points that bear on the interpretation of this situation are discussed further in later chapters. First, the core structures of mammalian glycoproteins built of mannose and GlcNAc are often the end products in simpler eukaryotes. In yeast and fungi, the core structures are extended rather than processed and they serve a structural role, forming the outer wall of these single-celled eukaryotes. Thus, the high mannose structure represents an evolutionary precursor to the more complex glycans and a pre-existing pathway was adapted to generate glycans with more complex terminal elaborations needed for recognition purposes in multicellular organisms. A second important aspect of the synthesis scheme is that an initially uniform core structure is present on the glycoproteins as they transit the early part of the secretory pathway and subsequent modification of this core occurs as they approach the cell surface. This spatial differentiation is likely to reflect the fact that the glycan portions of glycoproteins can serve different functions inside the cell, related to protein folding and intracellular trafficking, and outside the cell, where they may be involved in recognition events.

◉ See Chapters 9 and 10 for more on the biology of glycan recognition.

SUMMARY

N-linked glycans and the biosynthetic pathways that generate them are the cornerstone of our understanding of mammalian glycobiology. In spite of their diversity, all N-linked glycans on glycoproteins are generated through a common pathway, which starts in the endoplasmic reticulum with the transfer of a preformed oligosaccharide to an asparagine residue in the sequence Asn-X-Ser/Thr within a nascent polypeptide chain. The three major classes of N-linked oligosaccharides (high mannose, complex, and hybrid) result from the differential action of glycosidases and glycosyltransferases in the Golgi apparatus. The restricted specificities of these glycosidases and glycosyltransferases require that they act in a defined order. The structures found on a particular glycoprotein depend primarily on the glycosyltransferases being expressed in the cell type in which the glycoprotein is synthesized. Many glycoproteins are heterogeneous, because a variety of different oligosaccharides can be attached to a particular asparagine residue. In many of these cases, the individual oligosaccharide structures present may not have specific functions. However, as is illustrated in subsequent chapters, some very specific terminal elaborations can be added to the core structures to mediate specific functions in recognition and other processes. Understanding how selective glycosylation occurs by regulation of glycosyltransferase expression and activity remains a key objective for the future.

KEY REFERENCES

Abeijon, C. and Hirschberg, C.B. (1992). Topography of glycosylation reactions in the endoplasmic reticulum, *Trends in Biochemical Sciences* **17**, 32–36. This paper discusses the complexity of the glycosylation reactions taking place in the endoplasmic reticulum with particular reference to the need to translocate both proteins and donor sugars from the cytoplasm.

Burda, P. and Aebi, M. (1999). The dolichol pathway of N-linked glycosylation, *Biochimica et Biophysica Acta* **1426**, 239–257. This paper gives a detailed overview of the biosynthesis of the lipid-linked precursor for N-linked oligosaccharides.

Greenwell, P. (1997). Blood group antigens: molecules seeking a function?, *Glycoconjugate Journal* **14**, 159–173. The potential roles of the carbohydrate structures that form blood group antigens are reviewed critically.

Knauer, R. and Lehle, L. (1999). The oligosaccharyltransferase complex from yeast, *Biochimica et Biophysica Acta* **1426**, 259–273. The biochemistry of the enzyme oligosaccharyltransferase is reviewed in detail.

Kornfeld, R. and Kornfeld, S. (1985). Assembly of asparagine-linked oligosaccharides, *Annual Review of Biochemistry* **54**, 631–664. A classic summary of the biosynthesis of N-linked glycans is presented.

Muramatsu, T. (2000). Essential roles of carbohydrate signals in development, immune response and tissue functions, as revealed by gene targeting, *Journal of Biochemistry (Tokyo)* **127**, 171–176. This paper provides an overview of the conclusions that can be drawn about the functions of oligosaccharides from the phenotypes of knockout mice.

Ruiz-Canada, C., Kelleher, D.J., and Gilmore R. (2009). Cotranslational and posttranslational N-glycosylation of polypeptides by distinct mammalian OST isoforms, *Cell* **136**, 272–283. The different functions of two forms of oligosaccharyltransferase in co- and post-translational glycosylation are explained.

Zielinska, D.F., Gnad, F., Wiśniewski, J.R., and Mann, M. (2010). Precision mapping of an in vivo N-glycoproteome reveals rigid topology and sequence constraints, *Cell* **141**, 897–907. A proteomic approach confirms the specificity of the *N*-glycosylation target sequence and the topological constraints on sites of glycosylation.

QUESTIONS

2.1   Describe the two types of reactions that take place during the processing of N-linked oligosaccharides. What are the similarities and differences, in terms of substrate requirements, between glycosidases and glycosyltransferases?

2.2   Explain why two different glucosidases and several different mannosidases are required to remove glucose and mannose residues during the processing of N-linked oligosaccharides.

2.3   Inhibitors of glycosidases and glycosyltransferases have been useful in studying the N-linked glycosylation pathway. What N-linked oligosaccharide structures would you expect to find on glycoproteins isolated from a cell line treated with the following inhibitors?

   a Castanospermine, an inhibitor of glucosidase I

   b Deoxymannojirimycin, an inhibitor of Golgi mannosidase I

   c Swainsonine, an inhibitor of Golgi mannosidase II

2.4   Knockout mice lacking enzymes involved in the N-linked glycosylation pathway have been useful in studying the functions of oligosaccharides. Knocking out GlcNAc transferase I is embryonic lethal with the embryos dying at day 9, whereas mice lacking the gene for ST6Gal sialyltransferase (the sialyltransferase that adds sialic acid in 2–6 linkage to galactose) are viable and fertile but have defects in their ability to mount an immune response. What oligosaccharide structures would you expect to find on glycoproteins isolated from these knockout mice? What conclusions can be drawn about the biological functions of different types of N-linked oligosaccharides from the phenotypes of these mice?

2.5   The cDNA sequence for a secreted protein shows that it contains two Asn-X-Ser/Thr motifs but nothing is known about the glycosylation of the protein. If you consider that either or both of these sites could be glycosylated and the oligosaccharides processed to either a Man_5 high mannose structure or to a sialylated complex bi-antennary structure, how many possible glycoforms of the protein would there be? Explain why it is likely that such a protein would actually have very many more glycoforms.

2.6   The iron-binding protein transferrin is synthesized in the liver and secreted into serum. Transferrin has two sites for N-linked glycosylation and analysis of the glycosylation of human serum transferrin shows that it contains a mixture of sialylated bi-antennary and tri-antennary complex oligosaccharides. If you expressed transferrin in a mammalian cell line, would you expect the glycosylation of the recombinant form of transferrin to be the same as the glycosylation of the native serum form? What factors might cause differences in glycosylation of the two forms?

O-Linked glycosylation

LEARNING OBJECTIVES

By the end of this chapter, students should understand:

1 The structure of mucin-type glycans and the ways that they are organized on polypeptides

2 The structures and biosynthesis of glycosaminoglycans and their attachment to core polypeptides in proteoglycans

3 The functions of mucins and proteoglycans in water retention and tissue structure

4 The unusual features of cytoplasmic and nuclear glycosylation

The properties of two groups of glycoproteins, mucins and proteoglycans, are dominated by the large number of O-linked sugars that they bear. Much of this chapter is devoted to a discussion of these two classes of proteins and how glycosylation endows them with physical properties appropriate for the important biological functions they perform. A wide variety of additional types of O-linked glycosylation have been recognized, including some that serve structural roles and others that have signalling functions. A selection of these further functions of O-linked glycans are also discussed.

3.1 Mucins are large, heavily O-glycosylated proteins that hold water

The primary purpose of many **mucins** is to retain water at surfaces that are exposed to the environment but are not sealed by moisture-impermeable layers as in the skin. Thus, mucins must be present at the surfaces of the digestive and genital tracts, and in the respiratory system. Salivary, gastric, intestinal, vaginal, and nasal mucins are characterized by different polypeptide backbone structures to which the sugars are attached. Addition of clusters of sialylated glycans to groups of serine and threonine residues results in regions of strong negative charge that give the mucins the capacity to bind large amounts of water. Because they are highly hydrated structures, the energetic cost of releasing water from mucins reduces the rate of evaporation. Mucin glycosylation is based on a relatively simple set of core structures in which *N*-acetylgalactosamine (GalNAc) is linked to the side chain of serine or threonine and bears a single β1-3-linked Gal residue to form a core 1 structure (Figure 3.1). Typical O-linked glycans attached to mucin polypeptides consist of a disialylated form of the core 1 structure.

There appear to be as many as 19 MUC genes encoding mucin core polypeptides in humans. In addition to aiding in water retention, the mucins serve as lubrication and also help to protect from invasion by micro-organisms. The organization of a mucin polypeptide reflects these various functions. The polypeptides are exceptionally long, containing as many as 10,000 amino acids, and can be membrane-bound or secreted. Each polypeptide contains tandem repeats of relatively simple amino acid sequences rich in serine and threonine that are potential attachments sites for the O-linked glycans (Figure 3.2). These tandem repeats differ in sequence between mucin types and between species, but are identical or nearly identical within a specific mucin gene in one species. There can be more than a hundred copies of the tandem repeats in mucin genes. The genes are often polymorphic, with different numbers of repeats observed in different individuals. Further O- and N-linked glycosylation occurs in regions of unique sequence outside the tandem repeats.

The overall sizes of the mucin molecules are increased by formation of disulphide-linked oligomers. Globular terminal domains in secreted mucin polypeptides direct formation of covalent complexes larger than 1 MDa (Figure 3.3). Dense glycosylation of the mucins leads to an extended polypeptide conformation, because of the need to accommodate the bulk of the glycans and because of the charge repulsion between the sialic acid residues. The large size and elongated dimensions of the glycosylated mucin oligomers give the hydrated glycoprotein viscoelastic properties that are

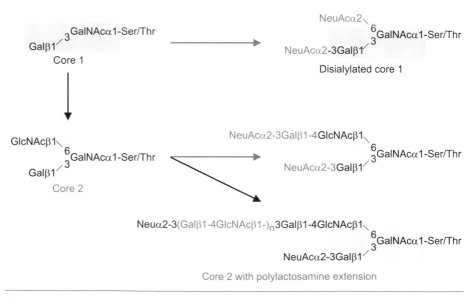

Figure 3.1 Core 1 and 2 structures attached to serine and threonine side chains through GalNAc residues. Such 'mucin-type' O-linked glycosylation is observed in mucins and in other glycoproteins.

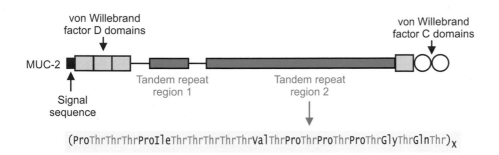

Figure 3.2 Overall organization of two mucins showing examples of tandem repeat sequences. Unique sequences flanking the tandem repeat domains specify secretion, membrane insertion, oligomerization, and bacterial binding functions. Glycosylated regions are shown in blue.

essential for the lubrication functions it performs. Secreted MUC-2 polypeptides are linked to each other through disulphide bonds to form branched structures. The long, hydrated, and cross-linked polymers can form gels such as those observed in nasal secretions.

Antibacterial properties of the mucins arise in two ways. The mechanical properties of viscous mucin solutions and gels serve to entrap potential pathogens so that they can be cleared from surfaces before they enter the body. In addition, some terminal domains have a specific affinity for bacterial cell surfaces, which can enhance entrapment and may also lead to immune responses against the bound organisms. Cell surface mucins can be integral membrane polypeptides, or they can associate with integral membrane polypeptides through terminal globular domains. In addition to providing functions similar to those of the secreted mucins, the cell surface mucins can regulate adhesion between cells.

◉ See section 13.8 for information on mucins and cancer.

3.2   Some cell surface proteins have mucin-like domains

Mucin-like domains occur in transmembrane proteins that also contain globular protein domains (Figure 3.3). These mucin-like domains are usually located between the membrane anchor and the globular domains, and serve to project the globular domains away from the surface of the membrane. An example of such an arrangement is found in the receptor for low-density lipoproteins. The ligand-binding domains located at the C-terminal end of this receptor are linked to the membrane by a mucin-like

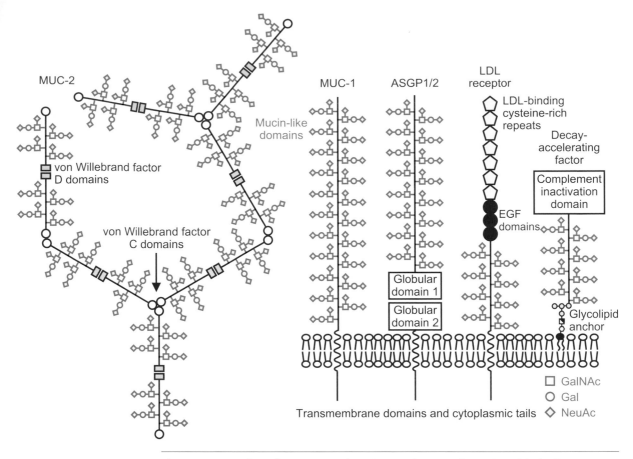

Figure 3.3 Comparison of secretory mucin MUC-2, membrane mucins MUC-1, ASGP (ascites sialoglycoprotein or rat MUC-4), and other membrane proteins containing mucin-like domains. The 'true mucins' are characterized by the extended stretches of O-glycosylated residues in tandem repeat sequences. Far more glycans than shown here are attached to the mucin-like domains (Figure 3.2). EGF, epidermal growth factor, LDL, low-density lipoprotein.

◯ See section 4.6 for more on glycolipid anchors.

domain containing multiple potential O-linked glycosylation sites. Extensive glycosylation of this type of mucin-like domain with disialylated core 1 structures generates a relatively rigid stalk, with the polypeptide maintained in an extended conformation by the bulky, charged glycans. Decay-accelerating factor is an example of a protein in which a globular domain is linked to a membrane by an elongated mucin-like domain and a glycolipid anchor.

Cell surface mucins can mediate cell adhesion as a result of specific interactions of protein receptors with the O-linked sugar structures. Many such interactions result from unique terminal elaborations attached to the ends of polylactosamine chains on core 2 structures (Figure 3.1). The core 2 structures are formed by derivatization of the protein-linked GalNAc with a β1-6-linked *N*-acetylglucosamine (GlcNAc) residue. As in N-linked glycans, extensions can be attached to core 2 in the form of a single lactosamine unit or a polylactosamine chain. Sialic acid is often attached to the branches of the core 2 structure and acts as a cap because the presence of sialic acid precludes addition of further units to the polylactosamine chain. The best understood examples of cell adhesion epitopes on O-linked glycans are ligands for the selectin cell adhesion molecules. These ligands are membrane mucins in which most of the O-linked sugars are simple core 1 structures that extend the polypeptide from the cell surface. Only a single attachment site, located near the end of the polypeptide that is furthest from the membrane, bears an elongated core 2 structure capped with the terminal structure needed for the adhesion event. This organization illustrates the dual role of O-linked sugars in conferring physical properties on polypeptide chains as well as presenting specific recognition tags.

⊙ See sections 9.5 and 9.6 for more on the selectins.

Glycophorin A was one of the first heavily O-glycosylated cell surface proteins to be identified. Sugar and protein in human erythrocyte glycophorin A combine to form antibody recognition sites. Differences in MN blood group reactivity are determined by the amino acid sequence at the extracellular N-terminus of glycophorin A, but reaction with either anti-M or anti-N antibodies requires attachment of sialic acid to O-linked glycans on adjacent serine and threonine residues.

3.3   Many soluble and cell surface glycoproteins contain small clusters of O-linked sugars

Individual glycosylated serine and threonine residues, and small clusters of mucin-type O-linked glycans, are found in soluble and cell surface glycoproteins. These glycosylation sites are characteristically located in hinge or linker regions between folded, globular domains in multidomain proteins. Examples include IgA (Figure 3.4), which is glycosylated in the hinge region between the Fab and Fc portions of the heavy chain, and the macrophage mannose receptor, which is glycosylated in several of the linker regions that separate globular carbohydrate-recognition domains. The exact function of O-linked sugars in linker regions is hard to establish, but it seems likely that they impart rigidity to otherwise flexible segments of polypeptide. These sugars may also provide protease resistance for segments of proteins that would be particularly vulnerable to digestion if they were not covered by sugars.

⊙ See sections 9.4, 10.8, and 10.9 for more on the mannose receptor.

3.4   Biosynthesis of mucin-type sugars occurs by sequential addition of monosaccharides to proteins in the Golgi apparatus

The O-glycosylation machinery utilizes glycosyltransferases analogous to those in the N-linked biosynthetic pathway, but the organization of these enzymes is substantially different. Addition of O-linked cores differs in two important respects from the addition of an N-linked core. First, all sugars are added one at a time in a step-wise series of reactions, starting with the first GalNAc residue attached to a serine or

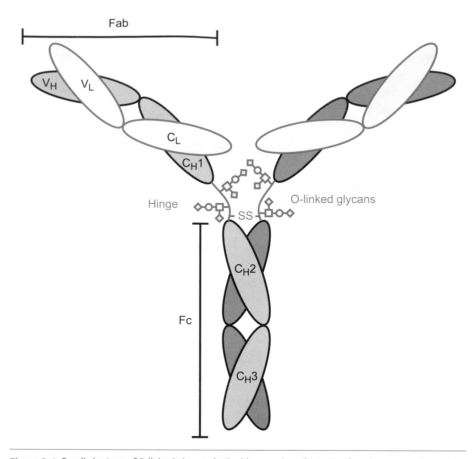

Figure 3.4 Small clusters of O-linked glycans in the hinge region of IgA. The five glycan attachment sites in each hinge may contribute to determining the conformation of the hinge and its resistance to proteolysis.

threonine residue. There is no preformed core or *en bloc* transfer. Second, there are no simple target sequences for O-linked glycosylation analogous to the Asn-X-Ser/Thr sequences that define N-linked glycosylation sites.

There are multiple reasons for the lack of a consensus O-glycosylation sequence. While all N-linked core glycosylation is catalysed by a single oligosaccharyltransferase, there are numerous transferases that can attach GalNAc to serine and threonine residues. These enzymes differ in their specificity for different sequences of amino acids surrounding the glycosylation target, so examination of the total pool of glycosylation sites does not reflect the specificity of any one enzyme. As individual enzymes have been cloned and expressed, it has become possible to look at the specificities of individual members of the family using synthetic peptides with different amino acids flanking potential glycosylation sites. Although such studies fail to reveal unique patterns of amino acids that determine glycosylation, rules can be discerned for individual enzymes. The target glycosylation sites tend to occur in regions rich in proline and alanine as well as serine and threonine residues. R-type lectin domains in some of the transferases bind to nearby O-linked sugars, targeting the catalytic domain to act on nearby serine and threonine residues. Enzymes that require previous glycosylation in order to act have the potential to work progressively, extending a glycosylated region that was initiated by another transferase, and thus generating clusters of glycosylated amino acid residues such as those in mucins.

◉ For more on R-type lectin domains see **section 11.6**.

Unlike the addition of the N-linked core, which takes place concomitant with or shortly after protein synthesis in the endoplasmic reticulum, the addition of core GalNAc residues to serine and threonine takes place post-translationally in the Golgi apparatus. Glycosyltransferases that add additional sugars to the O-linked structures are distributed through the Golgi apparatus, much like the enzymes that create the terminal elaborations of the N-linked structures. This intracellular distribution of the transferases, as well as the fact that all the sugars of the O-linked glycans are added as monosaccharides from nucleotide sugar donors, provides an interesting parallel to the terminal modification of the N-linked sugars. The comparison suggests that O-linked structures are somewhat like the terminal parts of N-linked sugars conjugated directly to proteins.

◉ See **section 5.6** for more on glycosytransferases in the Golgi apparatus.

3.5 Proteoglycans are heavily O-glycosylated proteins that give strength to the extracellular matrix

The second major class of heavily O-glycosylated proteins consists of the proteoglycans. Like mucins, proteoglycans bind water, but they provide structure rather than lubrication. Whereas each O-linked glycan of a mucin is small, often consisting of just a few sugar residues, the glycans attached to proteoglycans are very large, containing as many as a hundred residues. The monosaccharide units in these glycans are arranged in linear chains consisting of alternating residues of amino sugar derivatives and hexose derivatives (Figure 3.5). Thus, each structure can be described in terms of an underlying repeating disaccharide unit. The names of these glycosaminoglycans reflect tissues from which they were originally isolated. For example, hyaluronic acid, chondroitin sulphate, dermatan sulphate, and keratan sulphate are derived from hyaline membranes, cartilage, and skin (Figure 3.5).

Proteoglycans fall into two major classes: those found in extracellular matrix and those located in plasma membranes. The membrane proteoglycans are discussed in detail in Chapter 12. The best-understood matrix proteoglycans are those found in the structural tissues, particularly cartilage. While the collagen fibres of cartilage provide stiffness (resistance to bending) and strength (resistance to pulling), proteoglycans provide resilience, which is resistance to compression under pressure. This function depends on a high degree of super-molecular organization. The protein core of the major cartilage proteoglycan, aggrecan, organizes the glycosaminoglycans that are attached at intervals along the polypeptide (Figure 3.6). The attachment region contains a serine–glycine sequence conjugated to a xylose residue. This core is

Glycosaminoglycan	A unit	B unit	Protein core	Linkage	Tissues
Hyaluronic acid	GlcA	GlcNAc	No	None	Connective tissues Skin Cartilage Synovial fluid
Chondroitin sulphate	GlcA	GalNAc	Yes	O-Xylose	Cartilage Cornea Bone Skin Arteries
Dermatan sulphate	GlcA/IdoA	GalNAc	Yes	O-Xylose	Skin Blood vessels Heart valves
Heparan sulphate	GlcA/IdoA	GlcNAc	Yes	O-Xylose	Lung Arteries Cell surfaces
Keratan sulphate	Gal	GlcNAc	Yes	N-GlcNAc	Cartilage Cornea

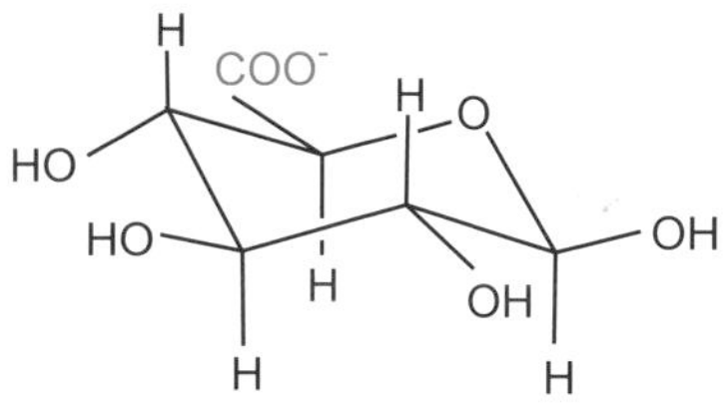

D-Glucuronic acid
(GlcA)

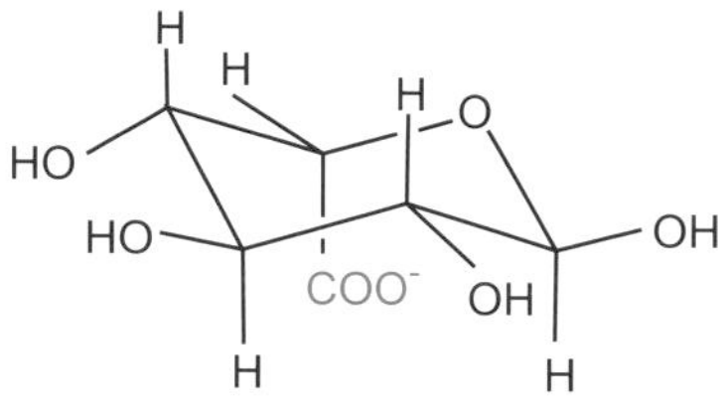

L-Iduronic acid
(IdoA)

Galβ1-3Galβ1-4Xylβ1-Ser
|
Gly

O-Xylose linkage region

Figure 3.5 Glycosaminoglycan structures and typical sites of expression. Keratan sulphate, derived from polylactosamine repeats, is elaborated on a typical N-linked core structure, while the other protein-bound glycosaminoglycans are attached to serine residues through a special linkage region.

then extended by a chondroitin sulphate chain. In addition, keratan sulphate chains are elaborated on typical N-linked glycans as they represent a sulphated form of the polylactosamine extensions. The mass of the covalent unit consisting of protein and attached glycans reaches as much as 2.5×10^6 Da, which is approximately the size of a ribosome.

Flanking the extended, O-glycosylated central portion of the aggrecan polypeptide are globular domains. Two domains at the N-terminal end of the polypeptide, designated G1 and G2, consist of repeating motifs that are homologous to the link protein, a small polypeptide that associates with this end of aggrecan. Both the link protein and the G1 module anchor multiple aggrecan polypeptides to chains of hyaluronic acid. The resulting aggregates can reach molecular weights of 2.5×10^8. The C-terminal G3 domain of aggrecan contains epidermal growth factor-like, complement control-type, and C-type lectin-like domains. The C-type lectin domain

◐ See section 9.1 for more on C-type lectin domains.

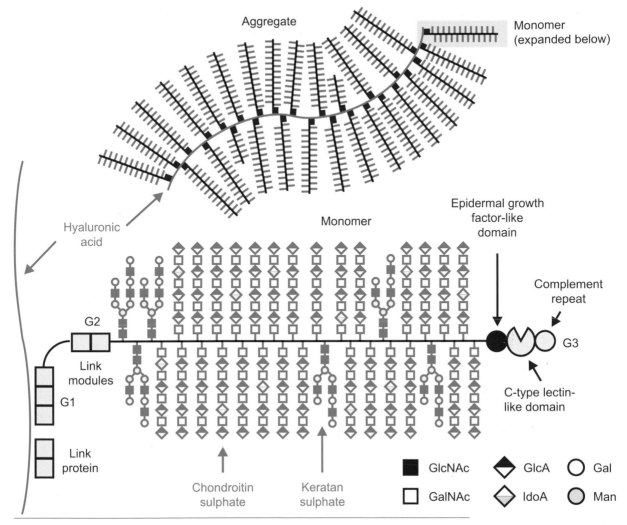

Figure 3.6 Organization of the aggrecan polypeptide and assembly of the polypeptide into larger aggregates. Approximately a 100 O-linked chondroitin sulphate chains, each containing about 100 sugar residues, and 30 N-linked keratan sulphate chains, each containing up to 50 sugar residues, are attached to the core polypeptide. Between 100 and 200 core polypeptides associate with a single hyaluronic acid molecule in the aggregate.

can bind sugars and proteins in the matrix. The G3 domain is present when the polypeptide is synthesized, but is often removed by proteolytic processing as tissue ages. These results suggest that this portion of the core polypeptide may play a transient role in cartilage organization.

The ability of proteoglycan aggregates to provide resilience in cartilage is a consequence of their highly hydrated state. The binding of water is mediated by the sugars as well as by the large number of sulphate residues attached to the glycosaminoglycans. As a result of this hydration, the volume occupied by an aggregate is very large and it can only be reduced, under pressure, by the thermodynamically unfavourable release of bound water. The aggregates act as molecular shock absorbers or very stiff sponges, releasing water slowly under pressure and taking it back up when pressure is released. Cartilage in various joints and in vertebral discs effectively cushions the jolts that would otherwise be transmitted by activities such as walking. When different species such as rats, humans, and cows are compared, the glycosaminoglycan attachment regions are found to be longer in the larger animals, because segments of the aggrecan gene encoding this portion of the polypeptide are expanded. Increased size of proteoglycan aggregates is needed to deal with increased loads on the joints of larger animals. The importance of cartilage proteoglycans is demonstrated by mutations in the core proteins and in the enzymes that glycosylate them. In chickens, the naturally occurring nanomelia mutation results in a truncated aggrecan core protein lacking much of the glycosaminoglycan attachment region and the C-terminal G3 globular domain. The mutant protein fails to support normal long bone development, resulting in drastic skeletal abnormalities.

Proteoglycans are also found in extracellular matrices of tissues other than cartilage. The core proteins of neurocan in brain and versican in blood vessels, skin, and other tissues closely resemble aggrecan. However, there are additional core proteins that can be conjugated with a variety of different glycosaminoglycan chains. One of the most abundant is perlican, which is a complex, modular protein consisting of multiple copies of at least six different types of globular domains and bearing just a few glycosaminoglycan chains. We do not yet have a detailed understanding of how different core protein architectures and certain selections of glycosaminoglycans provide physical and organizational properties that are particularly well-suited to the requirements of individual tissues.

3.6 Biosynthesis of proteoglycans requires several modifying enzymes in addition to glycosyltransferases

The target sequences for attachment of glycosaminoglycan chains are generally serine–glycine sequences. However, not all serine–glycine sequences are derivatized in this way, so there must be additional signals in the protein structure that define attachment sites. Addition of chains in clusters results from processive activity of the core xylosyltransferase, in which adjacent serine–glycine sequences become targets for the enzyme after glycosylation of an initial sequence. Although there is no consensus sequence for xylose addition, a cluster of negative charges to one side of the serine–glycine sequence often directs the enzyme to start. As most species have a single xylosyltransferase, initiation and propagation must be performed by the same enzyme.

Like mucin-type glycans, proteoglycans are assembled by successive one-step additions of monosaccharides. After assembly of the common trisaccharide core structure, the attachment of different glycosaminoglycans to different proteoglycan core proteins probably reflects, at least in part, different patterns of transferase expression in different tissues. In addition, experiments with synthetic acceptor substrates indicate that the pattern of flanking amino acids can influence the type of chain to be synthesized. However, the rules by which this information is decoded remain to be established.

Elongation of glycosaminoglycan chains involves sequential addition of alternating amino sugar and sugar acid residues. The glucuronic acid and GlcNAc-transferases that synthesize heparan form a complex that can elongate the growing chain in a progressive manner. Similar complexes are probably also involved in the synthesis of other types of glycosaminoglycans. It is less clear what determines the ultimate size of the glycosaminoglycan chains. Unlike the polylactosamine chains of N-linked and mucin-type O-linked oligosaccharides, there does not appear to be a competing capping reaction. The balance between enzymes that initiate the repeat structure and those that elongate the chains can definitely affect chain length. Other factors that may limit the size of the chains are overall residence time in the appropriate compartments of the Golgi apparatus, or the physical dimensions of the cisternae.

The final structures of most proteoglycans are the result of a further series of modifications to the initially synthesized glycosaminoglycan chains. These modifications establish sequences involved in growth factor and fibronectin binding. This process is again an untemplated scheme in which each reaction is carried out separately, although the modifications mostly proceed in an obligatory order (Figure 3.7). During heparan sulphate biosynthesis, the first step is *N*-deacetylation followed by *N*-sulphation. These steps are linked to each other because both enzymatic activities reside in a single polypeptide chain. *N*-sulphation is followed by 5-epimerization of glucuronic acid residues to iduronic acid. Finally, additional sulphotransferases attach sulphate to the 6 position of *N*-sulphated GlcNAc and the 2 position of iduronic acid residues. The initial *N*-deacetylation reaction lays down the pattern for subsequent sulphation. The requirement of subsequent enzymes for previously modified substrates explains how clusters of modified sites, known as S-domains, are formed in the glycosaminoglycans. In addition to the intracellular processing, proteoglycans can be further modified by sulphatases after secretion or presentation at the cell surface.

Altogether, the biosynthesis of a fully glycosylated proteoglycan molecule can involve tens of thousands of individual reactions. The final structures are heterogeneous in detail, but reproducibly contain the structural elements needed for proteoglycan function. The presence of key structures at appropriate density is a result of a combination of additions and modifications directed by glycosyltransferases and other enzymes that require specific peptide and carbohydrate sequences to act.

➲ See sections 12.1–12.3 for more on proteoglycans and growth factors.

3.7   Unusual types of O-linked glycosylation are found on some proteins

Much of the variety in O-linked glycosylation comes from different terminal elaborations attached to the core structures, particularly to the mucin core 2 structure. However, there are several other linkages between saccharides and hydroxyl groups

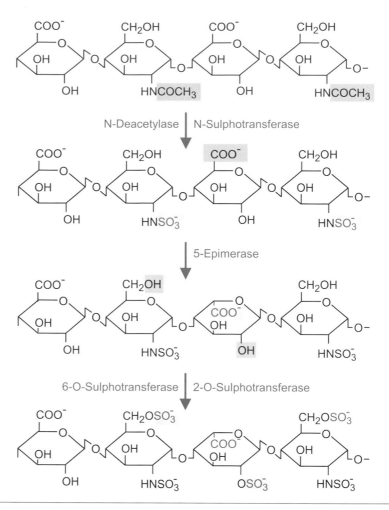

Figure 3.7 Biosynthesis of heparan sulphate requires assembly of the repeating backbone structure and replacement of the *N*-acetyl groups with *N*-sulphate groups. This is mediated by a bifunctional enzyme. Selective epimerization of glucuronic acid residue to iduronic acid, followed by 6-*O*-sulphation of GlcNAc residues and 2-*O*-sulphation of iduronic acid residues occurs in an obligatory sequence. Selective modification of certain regions within extended heparan sulphate chains produces unique sequences such as the one shown, which is able to bind with high affinity to fibroblast growth factors.

on amino acid side chains. In addition to GalNAc and fucose, mannose is also sometimes linked to serine or threonine residues in cell surface and secreted proteins. In a few instances, sugars are attached to the hydroxyl group of tyrosine residues. For example, a glucosyl tyrosine linkage is found in the core region of glycogen which serves as the anchor for growth of the branched glycogen structure.

One of the most abundant types of *O*-glycosylation is found in collagens and proteins with collagen-like triple helical domains. Within these domains, lysine residues found at the Y position of Gly-X-Y triplets are usually modified by lysyl hydroxylase,

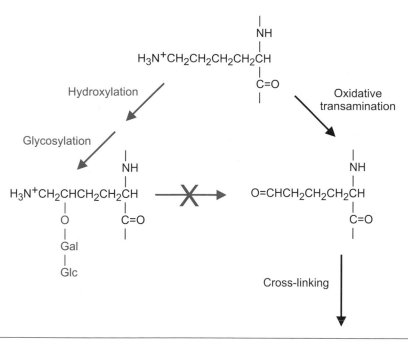

Figure 3.8 Alternative modifications of lysine residues in collagen. Addition of a disaccharide to hydroxylysine residues is believed to hinder formation of cross-links between collagen molecules.

which generates 4-hydroxylysine (Figure 3.8). Many of these 4-hydroxylysine residues are conjugated with a Glcα1-2Galβ disaccharide. Inhibition of lysine hydroxylation, and hence of glycosylation, does not appear to affect triple helix formation, but it does affect the cross-linking of fibres that are formed by bundling of triple helical domains. The effect may be a result of a direct role of the sugars in the packing arrangement within fibres. Alternatively, it may reflect competition between two possible modifications of the lysine side chains, because hydroxylation and glycosylation appear to be mutually exclusive with oxidation to form aldehyde groups that can cross-link fibres by forming Schiff bases with other lysine residues.

3.8 Cytoplasmic and nuclear proteins can be modified by addition of O-linked *N*-acetylglucosamine

Although most glycosylation of proteins occurs in extracytoplasmic compartments, O-linked GlcNAc (*O*-GlcNAc) is quite common in the cytoplasm and nucleus. A soluble GlcNAc-transferase catalyses attachment of GlcNAc in β1 linkage to serine and threonine residues. In addition to being unusual in its intracellular localization, *O*-GlcNAc is also almost unique among glycosylations in being readily reversible. A soluble *N*-acetylglucosaminidase removes the GlcNAc, which can then be replaced by the transferase (Figure 3.9). The extent of GlcNAc modification of

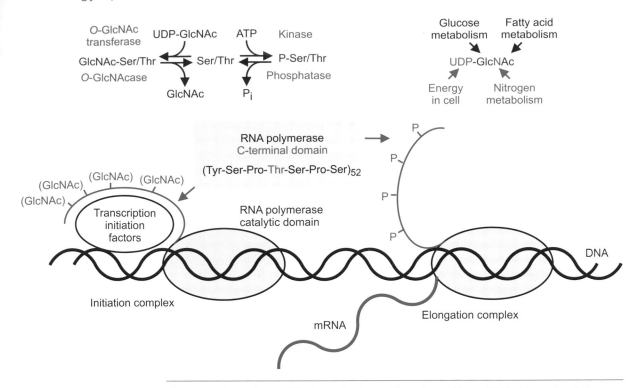

Figure 3.9 Alternative modifications of serine and threonine residues with GlcNAc or phosphate. Actions of kinases and phosphatases can be reciprocal to enzymes that add or remove GlcNAc. Portions of UDP-GlcNAc may respond to different aspects of cell metabolism. Phosphorylation of the extended C-terminal domain of RNA polymerase results in dissociation of transcription initiation factors and conversion to the elongation complex. ATP, adenosine triphosphate; UDP, uridine diphosphate. Adapted in part from: Wells, L., Vosseller, K. and Hart, G.W. (2003). A role for N-acetylglucosamine as a nutrient sensor and mediator of insulin resistance, *Cellular and Molecular Life Sciences* **60**, 222–228.

any particular serine or threonine residue thus reflects a dynamic balance between the actions of the transferase and the glycosidase.

The list of proteins that are targets for attachment of O-GlcNAc is long and ever growing. Many of the targets first identified were associated with the nucleus and the nuclear pore complex, including many transcription factors. However, examples of proteins bearing O-GlcNAc now encompass virtually every category of protein exposed to the cytoplasm, including proteins of the cytoskeleton and the plasma membrane in addition to soluble proteins. O-GlcNAc has been identified in all eukaryotic cells, including single-celled organisms such as yeast.

The dynamic and widespread nature of the O-GlcNAc modification is reminiscent of intracellular protein phosphorylation. Many proteins that are phosphorylated can also be modified with O-GlcNAc. The nature of target sequences for O-GlcNAc addition suggests that the specificity of this modification may resemble the specificity of certain protein kinases. Although there is no simple consensus sequence that serves as a target for the GlcNAc-transferase, comparison of naturally modified proteins and synthetic peptide substrates reveals a preference for proline and alanine in addition to serine and threonine residues in target sequences. Interestingly, this distribution is actually somewhat like the distribution around sites of GalNAc addition to extracellular glycoproteins. Unlike the protein kinases and phosphatases, which are very large families, there seems to be only one type of cytoplasmic GlcNAc-transferase and *N*-acetylglucosaminidase in a given organism.

Identifying an effect of GlcNAc addition on the function of intracellular proteins has proved difficult. It has been proposed that GlcNAc addition may serve to modulate phosphorylation, because addition of GlcNAc to serine or threonine could block phosphorylation of these residues. The potential for two types of modification of a single target amino acid makes possible additional levels of control. For example, phosphorylation of specific sites would require the activity both of the *N*-acetylglucosaminidase and a kinase. It has also proven difficult to correlate the level of O-linked GlcNAc on particular proteins, or the activity of the GlcNAc-transferase or the *N*-acetylglucosaminidase, with any type of signalling mechanism. One of the most tantalizing examples in this regard is the extended C-terminal domain of RNA polymerase II, the enzyme that transcribes messenger RNA molecules (Figure 3.9). The extended terminal domain consists of 52 copies of a seven-amino acid sequence that is a target for both the GlcNAc-transferase and for protein kinases. The unphosphorylated form of the terminal domain interacts with transcription factors in the transcription initiation complex and phosphorylation results in release of the enzyme from these factors and transformation to the elongation complex. Thus, it is plausible that O-GlcNAc attached to the terminal domain might regulate phosphorylation, and hence conversion from the initiation to the elongation form of the enzyme.

3.9   O-linked *N*-acetylglucosamine is part of a metabolic sensor system that is affected in diabetes

In at least some cells, the level of O-GlcNAc attached to proteins is modulated by glucose concentration. Increasing extracellular glucose concentration increases the modification of intracellular proteins with O-GlcNAc. The link between glucose concentration and O-GlcNAc addition is UDP-GlcNAc, the activated sugar donor used by the O-GlcNAc-transferase. UDP-GlcNAc is synthesized via the hexosamine biosynthetic pathway. In this pathway, fructose 6-phosphate derived from glucose is first converted to glucosamine 6-phosphate by transfer of the amido group from glutamine followed by isomerization in a reaction catalysed by glutamine:fructose 6-phosphate amidotransferase. Glucosamine 6-phosphate is converted through a series of steps to UDP-GlcNAc for use in glycosylation reactions. O-GlcNAc-transferase activity is particularly sensitive to changes in UDP-GlcNAc concentration, making O-GlcNAc addition also sensitive to changes in glucose concentration. This pathway may respond to cellular metabolism in other ways as well (Figure 3.9).

Increased glucose flux through the hexosamine biosynthetic pathway in insulin-sensitive tissues has been linked to the development of insulin resistance that causes type 2 diabetes. Evidence that this pathway acts as a sensor of the nutrient status of the cell and regulates insulin activity includes the finding that overexpression of glutamine:fructose 6-phosphate amidotransferase in skeletal muscle, pancreatic β-cells, and adipose tissue in mice leads to insulin resistance. Targeted overexpression of O-GlcNAc-transferase in muscle and adipose tissue also causes insulin resistance. Overexpression of each of these two enzymes leads to increased modification of proteins by O-GlcNAc, which could therefore be the mechanism underlying insulin resistance that causes type II diabetes. Many proteins involved in insulin signalling pathways, including insulin receptor substrate 1 (IRS-1), are modified by O-GlcNAc but whether this modification affects activity is not known. Activity of glycogen synthase, the enzyme responsible for synthesis of glycogen from glucose in response to insulin stimulation, is affected by O-GlcNAc addition. In the presence of high glucose, more O-GlcNAc is added to glycogen synthase and it becomes less sensitive to activation by protein phosphatase 1 in response to insulin. Thus, increased O-GlcNAc modification of glycogen synthase is likely to be one of the mechanisms contributing to insulin resistance.

As well as being relevant to the pathogenesis of diabetes, the link between extra-cellular glucose concentration and modification of intracellular proteins with O-GlcNAc might be part of a more general response of cells to the availability of nutrients. Addition of O-GlcNAc to transcription factors or to proteins involved in signalling pathways could modulate cellular responses to the environment. In addition, modification of the proteasome by O-GlcNAc inhibits protein degradation, providing a mechanism for linking the availability of proteins and amino acids to the nutritional state of the cell.

Cells lacking the GlcNAc-transferase for making O-GlcNAc are not viable, indicating that this modification plays an essential part in cell physiology. Galactosyltransferase introduced into cells causes modification of the O-GlcNAc with a galactose residue and prevents its release by the N-acetylglucosaminidase. These cells die within one cell cycle, confirming that at least one essential role of O-GlcNAc is in control of the cell cycle. Thus, the evidence to date suggests that O-GlcNAc plays a vital and unique role in the control of cellular processes, but there are important missing links in our understanding of the mechanisms of O-GlcNAc function.

SUMMARY

O-linked glycans are extremely diverse in both structure and function, and the full extent of this diversity has not yet been established. N-linked glycans share at least a common protein–glycan linkage and fall into relatively few structural classes, reflecting a common biosynthetic pathway that diverges in the late stages. In contrast, the groups of O-linked glycans are built on different protein–glycan linkages, in which GalNAc, fucose, GlcNAc, mannose, xylose, or galactose can be attached to serine, threonine, or hydroxylysine residues. Terminal structures on some of the O-linked glycans are similar or identical to structures on N-linked glycans, suggesting that there may be functional overlap between these types of glycosylation. Indeed, enzymes involved in the synthesis of these structures may be shared, indicating that they have co-evolved to display similar terminal elaborations that can be involved in various recognition events. However, O-linked glycans have also evolved to serve unique functions, many of which remain to be explored.

KEY REFERENCES

Butkinaree, C., Park, K., and Hart, G.W. (2010), O-linked β-N-acetylglucosamine (O-GlcNAc): Extensive crosstalk with phosphorylation to regulate signalling and transcription in response to nutrients and stress, *Biochimica et Biophysica Acta* **1800**, 96–106. Many of the proposed functions of O-GlcNAc modification are summarized in part of an entire issue of this journal devoted to the subject.

Esko, J.D. and Selleck, S.B. (2002). Order out of chaos: assembly of ligand binding sites in heparan sulfate, *Annual Review of Biochemistry* **71**, 435–471. The biochemistry and biology of heparan sulphate are reviewed in detail.

Hanisch, F.-G., Reis, C.A., Clausen, H., and Paulsen, H. (2001). Evidence for glycosylation-dependent activities of polypeptide N-acetylgalactosaminyltransferases rGalNAc-T2 and -T4 on mucin glycopeptides, *Glycobiology* **11**, 731–740. Experiments demonstrating specific substrate requirements for O-glycosylation by GalNAc-transferases are presented.

Heikkinen, J., Risteli, M., Wang, C., Latvala, J., Rossi, M., Valtavaara, M., and Myllyla, R. (2000). Lysyl hydroxylase 3 is a multifunctional protein possessing collagen glucosyltransferase activity, *Journal of Biological Chemistry* **275**, 36158–36163. Characterization of the enzyme that hydroxylates and glucosylates lysine residues in collagen is presented.

Kim, Y.S., Gum, J., Jr., and Brockhausen, I. (1996). Mucin glycoproteins in neoplasia, *Glycoconjugate Journal* **13**, 693–707. Changes in the expression and structure of mucins in cancer are discussed.

Raman, J., Fritz, T.A., Gerken, T.A., Jamison, O., Live, D., Liu, M., and Tabak, L.A. (2008). The catalytic and lectin domains of UDP-GalNAc:polypeptide α-N-acetylgalactosaminyltransferase function in concert to direct glycosylation site selection, *Journal of Biological Chemistry* **283**, 22942–22951. This article summarizes some of the differences between the multiple enzymes that initiate mucin-type glycosylation and suggests a role for lectin domains in targeting the catalytic domains.

Thornton, D.J., Rousseau, K., and McGuckin, M.A. (2008). Structure and function of the polymeric mucins in airway mucus, *Annual Review of Physiology* **70**, 459–486. Although focused on the airway mucins, this review provides a good overview of all of the mucin glycoproteins and their physical properties.

Van den Steen, P., Rudd, P.M., Dwek, R.A., and Opdenakker, G. (1998). Concepts and principles of O-linked glycosylation, *Critical Reviews in Biochemistry and Molecular Biology* **33**, 151–208.

The structures, biosynthesis, and functions of different types of O-linked glycans are reviewed.

Vertel, B.M. (1995). The ins and outs of aggrecan, *Trends in Cell Biology* **5**, 458–464. The biology and biochemistry of the proteoglycan aggrecan are reviewed.

Vestweber, D. and Blanks, J.E. (1999). Mechanisms that regulate the function of selectins and their ligands, *Physiological Reviews* **79**, 181–213. The structures and functions of the selectins and their glycoprotein ligands are reviewed in detail.

QUESTIONS

3.1   What are the similarities and differences between the pathways of O-linked glycosylation and N-linked glycosylation?

3.2   What are the main functions of O-linked glycans? What functions do O-linked glycans have in common with N-linked glycans?

3.3   In what ways does the modification of proteins with O-linked GlcNAc differ from other types of O-linked glycosylation and N-linked glycosylation? What unusual features of this type of glycosylation have led to the suggestion that it might modify the activities of proteins as part of a signalling system?

3.4   Discuss the conflicting evidence for the role of *O*-GlcNAc based on use of different inhibitors of *O*-GlcNAcase.

Reference: Macauley, M.S., Bubb, A.K., Martinez-Fleites, C., Davies, G.J., and Vocadlo, D.J. (2008). Elevation of global *O*-GlcNAc levels in 3T3-L1 adipocytes by selective inhibition of *O*-GlcNAcase does not induce insulin resistance, *Journal of Biological Chemistry* **283**, 34687–34695.

3.5   Why is it difficult to predict sites of O-linked glycosylation from examination of the amino acid sequences of proteins? What can be learned from comparison of sequences of proteins known to have O-linked glycosylation sites?

Reference: Hema Thanka Christlet, T. and Veluraja, K. (2001). Database analysis of *O*-glycosylation sites in proteins, *Biophysical Journal* **80**, 952–960.

3.6   Compare and contrast the structures, biosynthesis, and functions of proteoglycans and mucins.

3.7   Assess the evidence that addition of O-linked fucose to proteins occurs in the endoplasmic reticulum rather than in the Golgi apparatus.

Reference: Luo, Y. and Haltiwanger, R.S. (2005). O-fucosylation of notch occurs in the endoplasmic reticulum, *Journal of Biological Chemistry* **280**, 11289–11294.

Glycolipids and membrane protein glycosylation

LEARNING OBJECTIVES

By the end of this chapter, students should understand:

1   Patterns of glycosylation of membrane glycoproteins

2   The structures of the different classes of glycolipids

3   The effects of deficiencies in breakdown of glycolipids

4   The structures of glycolipid anchors attached to proteins and their functions in different organisms

Animal cell surfaces are rich in glycoconjugates. Glycans displayed on glycoproteins and glycolipids of the plasma membrane create a meshwork that projects up to 100 Å from the surface (Figure 4.1). This meshwork, called the glycocalyx, can be visualized in thin section electron microscopy. The glycocalyx represents the outermost surface of the cell and plays a defining role in interactions between cells. Like the glycans attached to soluble glycoproteins, those found in the plasma membrane can serve both informational and structural roles, as discussed in this chapter.

4.1   Most integral membrane proteins are glycosylated

Some forms of membrane protein glycosylation have been discussed in previous chapters, but it is worth highlighting specific features common to membrane glycoproteins. Water-soluble proteins associated with the surface of cells are usually regarded as part of the extracellular matrix rather than part of the plasma membrane. Thus, the major glycoproteins of the plasma membrane are integral membrane proteins. In general, N- and O-linked glycans attached to these proteins are similar to those found on secreted glycoproteins, and the machinery used to synthesize and attach the glycans is common to both membrane and secreted proteins. However, some structures, such as those containing polylactosamine extensions, are more common on membrane proteins.

Figure 4.1 Electron micrograph of the glycocalyx at the surface of an erythrocyte. Reproduced with permission of the publisher from: Voet, D. and Voet, J. G. (1995). *Biochemistry*. Chichester: Wiley. Original was by courtesy of Harrison Latta, UCLA, with permission.

The distribution of glycosylation sites relative to the plasma membrane surface does not follow any fixed rules, but there are certain common patterns (Figure 4.2). In membrane proteins that have a single membrane-spanning sequence, it is quite common for at least one site of *N*-glycosylation to be found in the portion of the polypeptide closest to the membrane. Particularly in oligomeric membrane proteins, glycosylation in this region may form a collar around the stalks that project other domains away from the cell surface. A ring of oligosaccharides might help to orientate such proteins perpendicular to the surface of the membrane. Extended regions of O-linked glycosylation in integral membrane proteins have a similar role. The position of these glycans relatively near to the cell surface rather than at the surface of the glycocalyx makes them less likely to be targets for recognition events.

Membrane proteins that contain multiple membrane-spanning segments also frequently bear N-linked glycosylation. As a general rule, there is only a single glycosylation site per polypeptide and this is usually found in a relatively large loop containing 30 or more amino acids. Glycosylation often occurs in the most N-terminal extracellular loop of appropriate size in the protein. The selection of only a single site of glycosylation per polypeptide may reflect the limited amount of space available in the surface area above such proteins, in which the membrane domains are closely packed α helices. The sugars thus form a local umbrella over the protein.

4.2 Membranes contain glycolipids as well as glycoproteins

In addition to being presented at the surface of animal cells by attachment to integral membrane proteins, glycans are also attached to the head groups of lipids. The lipid portions of glycolipids, which are embedded in the membrane bilayer, fall into two structural categories (Figure 4.3). Glycolipids built on ceramide are known as

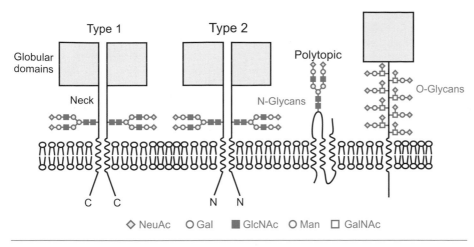

Figure 4.2 Summary of some of the common patterns of membrane protein glycosylation. Monotopic transmembrane proteins contain single membrane-spanning sequences in either of two orientations. Polytopic membrane proteins have multiple membrane-spanning sequences in various arrangements. Many membrane proteins bear both N- and O-linked glycans.

glycosphingolipids, because ceramide is formed by attachment of a fatty acid in amide linkage to the long-chain amino alcohol sphingosine. Glycosphingolipids can function like membrane glycoproteins in the presentation of potential recognition markers involved in cell–cell interactions. Glycosphingolipids also function in the organization of specialized membrane domains. In contrast, glycophospholipids built on a phosphatidylglycerol core provide a mechanism for anchoring proteins to the cell surface.

There are two main subgroups of glycosphingolipids that can be differentiated by whether the first sugar attached to sphingosine is galactose or glucose. The primary galactosphingolipid is the relatively simple sulphatide molecule, in which the galactose is sulphated on the 3 position. The head groups of the glucosphingolipids are generally more elaborate and often resemble the terminal glycans on glycoproteins.

The nomenclature applied to the glycosphingolipids can often be perplexing. The appellation 'sphingo', referring to the riddle of the sphinx, was applied to these glycolipids to reflect their unknown functions, but it could well be a comment on the nomenclature! Many of these lipids were isolated originally from the brain and the name of one of the common series of structures, the gangliosides, is based on this fact. The shortest members of the ganglio series, in which a single galactose residue is linked to the core glucose, are designated G3, addition of an *N*-acetylgalactosamine (GalNAc) residue generates the G2 members, and extension with a further galactose residue generates the G1 members. A second letter in the common abbreviations, which is inserted between the G and the number, reflects the state of sialylation of the core: M for monosialylated, D for disialylated, T for trisialylated, and so forth. Further lower case letters are added to distinguish isomers in which the sialic acid is attached at different positions.

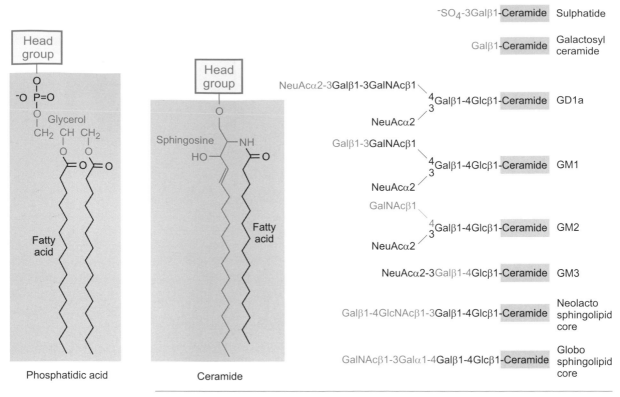

Figure 4.3 Covalent structures of the lipid portions of glycosphingolipids and glycophospholipids and the head groups of common glycosphingolipids. Glucosphingolipids are generally more complex than galactosphingolipids and can be built on at least six different backbone structures, including members of the ganglio (GM and GD), neolacto, and globo series shown.

At least seven other core sequences are also known. In the neolactoside series, the repeating unit of the core is the Galβ1-4GlcNAc disaccharide familiar from the poly-lactosamine structures on N- and O-linked glycans attached to glycoproteins. The similarity of these structures and of other structures in the cores and extensions of other glycolipid and glycoprotein glycans means that similar terminal elaborations can be built on to both types of glycoconjugates. Combining the diversity of such elaborations with the diversity of core structures means that there are hundreds of different species of glycolipids.

4.3   Glycosphingolipid biosynthesis occurs in the Golgi apparatus

The biosynthesis of glycosphingolipids resembles the biosynthesis of the dolichol-linked glycan that is transferred to asparagine residues in glycoproteins. The addition of the first glucose residue to initiate glucosphingolipid synthesis occurs from a nucleotide sugar donor on the cytoplasmic face of the endoplasmic reticulum membrane (Figure 4.4). Following a poorly understood transmembrane flipping process, further sugars are added after the lipid has moved through the secretory pathway so that it has access to nucleotide sugar donors in the Golgi apparatus. In contrast, the first step in galactosphingolipid biosynthesis takes place on the luminal side of the membrane, so no further flipping is required to access transferases for further additions. Glycosyltransferases involved in the first stages of glycolipid biosynthesis determine the spectrum of core structures present in any particular cell type. Some terminal structures are added by enzymes that are similar to those that create elaborations on N- and O-linked structures of glycoproteins. In most cases, however, there are distinct enzymes that modify the glycans of glycolipids and glycoproteins. The movement of glycolipids through the luminal compartments of the cell to the plasma membrane generally follows the pathway for membrane proteins.

4.4   Glycosphingolipids can generate distinct domains in the plasma membrane

Many glycolipids assemble into discrete micro-domains known as lipid rafts. The rafts, which persist once the glycolipids have reached the plasma membrane, probably result from the interaction of cholesterol with the sphingolipids within the lipid

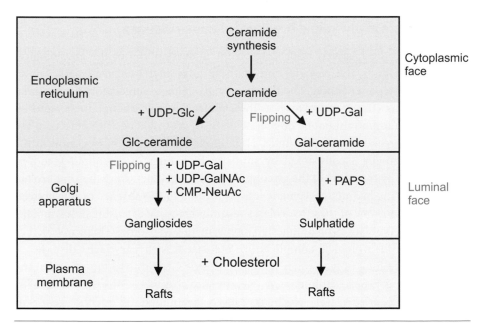

Figure 4.4 General pathways of glycosphingolipid biosynthesis. Both galactosyl- and glucosylsphingolipid synthesis is initiated on the cytoplasmic side of the membrane and is completed in the lumen, but the point at which transmembrane flipping occurs is different in the two pathways. The sulphate donor is 3′-phosphoadenosine 5′-phosphosulphate (PAPS). CMP, cytidine monophosphate; UDP, uridine diphosphate.

bilayer. In addition to the glycolipids, the sphingolipids include sphingomyelin, in which a phosphocholine head group replaces the glycan. The molecular interactions that underlie raft formation are poorly understood, but rafts are believed to form a gel phase distinct from the surrounding phospholipids which are in a fluid state. The rafts are resistant to solubilization with detergents under conditions that disperse the bulk phospholipids of the membrane. Distinct types of lipid rafts may have different glycolipid compositions. For example, the glycosphingolipid GM1 is a common component of many lipid rafts, but the related molecule GM3 is found only in rafts that are not associated with the protein caveolin. Caveolin is found on the cytoplasmic side of flask-shaped cell surface invaginations known as caveolae, which are involved in membrane trafficking events such as specialized forms of endocytosis.

Glycolipids modulate the behaviour of specific membrane receptors, at least in part through their ability to affect the localization of the receptors. For example, partition of the insulin receptor into a region of the membrane rich in ganglioside GM3 may remove the receptor from caveolae where it interacts with substrates that initiate intracellular signalling. The activity of the epidermal growth factor receptor is also modulated by GM3. In this case, the glycan headgroup of the glycolipid may directly interact with N-linked glycans on the receptor to inhibit receptor activation, possibly by preventing receptor dimerization.

4.5   Defects in glycolipid breakdown cause disease

Glycobiology and disease
Glycolipid storage disorders

⊙ See section 11.6 for more on bacterial toxins.

In contrast to the relatively poor state of our understanding of the physiological roles of glycolipids, the participation of these molecules in several pathological processes is well understood. The function of glycolipids as receptors for bacterial toxins serves as a useful model for how these molecules may function in recognition processes. An important family of genetic diseases is associated with defects in the pathways of breakdown of glycolipids. High levels of glycolipid biosynthesis in the brain are normally accompanied by similarly high rates of glycolipid turnover, which occurs by transport of the glycolipids to the lysosome where they are broken down by specific hydrolases. Genetic lesions leading to the absence or low activity of practically any one of these hydrolases result in blockage of the breakdown pathways and accumulation of undigested lipids in the lysosomes (Figure 4.5). For example, fatal neurological complications associated with Tay–Sachs disease are caused by the filling of lysosomes with ganglioside GM2 due to mutations in the gene encoding β-hexosaminidase A. Neurological complications are not usually seen in Gaucher disease, which is caused by mutations in the gene for β-glucocerebrosidase, the enzyme that degrades glucosylceramide. Symptoms of Gaucher disease include enlarged spleen and liver, and skeletal abnormalities due to accumulation of glucosylceramide in macrophages. Many Gaucher disease sufferers can be successfully treated by enzyme replacement therapy (Box 4.1). Fabry disease, caused by a defect in the enzyme α-galactosidase A, which removes the terminal

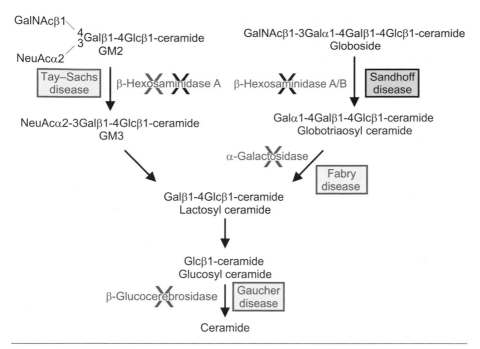

Figure 4.5 Lysosomal storage disorders resulting from defects in glycolipid breakdown. Loss of each enzyme causes build-up of the glycolipid that it would normally break down. In the case of Sandhoff disease, loss of both β-hexosaminidases A and B leads to build-up of both ganglioside GM2 and globoside.

galactose residue from globotriaosylceramide (also called ceramide trihexoside), is the only other glycosphingolipid storage disease in which lipid storage occurs mainly in peripheral tissues. In Fabry disease, globotriaosylceramide accumulation in the endothelial cells lining blood vessels eventually causes damage to organs such as the kidney and the heart.

4.6   Some proteins are attached to membranes through glycolipid anchors

Anchoring of proteins to membranes by covalent attachment to fatty acids and lipids is an important alternative to anchoring through hydrophobic amino acid sequences. In mammalian cells, more than a hundred membrane proteins are attached to the extracellular surface of the plasma membrane through glycolipids. Unlike the glyco-sphingolipids, these lipids are based on a **diacylglycerol** hydrophobic structure. Glycolipids attached to proteins are usually referred to as **glycosylphosphatidyl inositol (GPI) anchors**.

The structure of GPI anchors differs in different organisms but they share certain common features (Figure 4.6). The glycan that bridges between the protein and the

BOX 4.1   Glycotherapeutics: *Enzyme replacement, molecular chaperones, and inhibitors of glycolipid synthesis can be used to treat storage diseases*

The most common glycolipid storage disease, Gaucher disease, can be treated by enzyme replacement therapy in which the defective catabolic enzyme is replaced through injection of a purified active form of the enzyme. Enzyme replacement therapy is effective in many patients with Gaucher disease because, in this disorder, undigested glycolipid accumulates in lysosomes of macrophages in peripheral tissues but there is usually no accumulation in the central nervous system. Injected β-glucocerebrosidase reaches the macrophages in the bone marrow, spleen, and liver and breaks down accumulated glucosylceramide. Production of a recombinant form of β-glucocerebrosidase with glycans terminating in mannose residues allows the enzyme to be targeted to macrophage lysosomes through uptake by the macrophage mannose receptor (Chapter 10). Fabry disease is also now being treated by enzyme replacement therapy with a recently developed recombinant form of α-galactosidase A. Enzyme replacement therapy is not an effective treatment for the rare type 2 form of Gaucher disease or the other glycosphingolipid storage diseases where lipid accumulation in the brain causes fatal neurological damage, because the enzymes cannot cross the blood–brain barrier and thus do not reach the affected cells.

Another method of getting properly folded glucocerebrosidase to lysosomes is to facilitate folding of mutant forms of the enzymes so they are able to exit the endoplasmic reticulum and travel to the Golgi apparatus and on to the lysosomes. Many mutant forms of the protein are stabilized by substrate analogues that bind at the active site. These relatively simple compounds are often based on deoxynojirimycin, an amino sugar analogue of glucose. They are usually competitive inhibitors of the enzyme and can be made in cell-permeable form with butyl, nonyl, or other alkyl chains, so they can be taken orally, avoiding the need for the repeated intravenous injections required for enzyme replacement therapy, and they have the potential to reach enzyme in the brain. Similar inhibitors of enzymes defective in Tay–Sachs and other lysosomal storage diseases are also being developed.

D-Glucose   N-butyl deoxynojirimycin   N-nonyl deoxynojirimycin

A third approach to treatment of glycolipid storage disorders is to try to prevent or reduce accumulation of the glycolipids. In substrate deprivation, inhibitors are used to reduce the biosynthesis of glycosphingolipid so that smaller amounts reach the lysosomes. In many patients, the defects in the catabolic enzymes do not abolish activity completely and low residual activity may be sufficient to cope with degradation if the substrate level is kept low. Even if there is no residual enzyme activity, inhibition of glycolipid synthesis can slow down the accumulation of undegraded material so that toxic levels might not be reached. One inhibitor of glycolipid synthesis, N-butyl deoxynojirimycin, is now used as an alternative to enzyme replacement in treatment of some patients with Gaucher disease. *N*-butyl deoxynojirimycin inhibits the ceramide-specific glucosyltransferase that catalyses formation of glucosylceramide, the glycolipid that builds up in Gaucher disease. Glucosylceramide is a precursor in the biosynthesis of all glucosylsphingo-lipids, so inhibition of its synthesis reduces formation of the glycolipids that accumulate in Tay–Sachs disease, Sandhoff disease, and Fabry disease, and *N*-butyldeoxynojirimycin treatment is being tested in patients with Tay–Sachs disease. It is perhaps surprising that reduced synthesis of glycosphingolipids does not have physiological consequences, but Gaucher patients treated with *N*-butyldeoxynojirimycin experience no serious side effects.

Essay topic

- Compare and contrast the three different approaches for treating glycolipid storage disorders. Discuss how knowledge of glycobiology has been important in developing treatments for these diseases.

Lead references

- Brady, R.O. (2003). Enzyme replacement therapy: conception, chaos and culmination, *Philosophical Transactions of the Royal Society of London B*, **358**, 915–919.

- Ioannou, Y.A., Zeidner, K.M., Gordon, R.E., and Desnick, R.J. (2001). Fabry disease: Preclinical studies demonstrate the effectiveness of α-galactosidase A replacement in enzyme-deficient mice, *American Journal of Human Genetics* **68**, 14–25.

- Platt, F.M. and Butters, T.D. (2000). Substrate deprivation: A new therapeutic approach for the glycosphingolipid lysosomal storage diseases, *Expert Reviews in Molecular Medicine* **February 1**.

- Sawkar, A.R., Schmitz, M., Zimmer, K.-P., Reczek, D., Edmunds, T., Balch, W.E., and Kelly, J.W. (2006). Chemical chaperones and permissive temperatures alter the cellular localization of Gaucher disease associated glucocerebrosidase variants, *ACS Chemical Biology* **1**, 235–251.

lipid is attached to the **inositol** head group of the lipid by a glycosidic linkage of glucosamine (GlcN) to the 6 position. This is one of the few instances in which the free amino sugar, rather than its acetylated form, has been identified in glycoconjugates. All known lipid anchors contain a core of three mannose residues. A molecule of **ethanolamine** is linked to the third mannose residue by a phosphodiester bond and the exposed free amino group forms an amide bond with the carboxyl group of the C-terminal amino acid in the protein.

The basic GPI anchor is subject to various modifications. Points of diversity among different anchors include the length of the fatty acid tails attached to the glycerol moiety, the presence or absence of a second ethanolamine phosphate residue attached to the first mannose residue, and the presence or absence of a palmitate esterified directly to the inositol ring. GPI anchors were originally recognized because many proteins linked to membranes in this way can be released by treatment with phospholipase C. However, because palmitate esterified to the inositol ring inhibits this cleavage, sensitivity to phospholipase C cannot always be used as a diagnostic test for whether a protein is attached to a membrane through a GPI anchor.

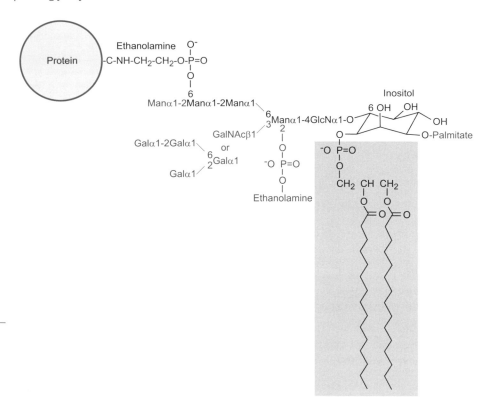

Figure 4.6 GPI anchor structure, with the common portions shown in black and variations found in different species shown in blue.

A further source of variability in the structures of GPI anchors is additional elaboration of the glycan. These extensions can take the form of a single GalNAc residue linked to the first mannose residue or a branched cluster of galactose residues. Thus, the overall organization of the glycan resembles complex N-linked oligosaccharides, with an inner core structure consisting of GlcN(Ac) and mannose, and exposed outer branches containing galactose and GalNAc.

4.7 Glycolipid anchors are added to proteins in the endoplasmic reticulum

The attachment of glycolipid anchors to proteins is analogous to the attachment of N-linked sugars because the anchor is pre-assembled as a unit before it is transferred to the protein. Like the dolichol lipid core structure in N-linked glycosylation, the construction of the core inositol lipid takes place in two topologically distinct stages (Figure 4.7). The addition of a GlcNAc residue, removal of the *N*-acetyl group and in some cases the further attachment of palmitate to the inositol ring all occur on the cytoplasmic face of the endoplasmic reticulum membrane. This structure is then transferred to the luminal side of the membrane where additions of mannose and ethanolamine phosphate take place. The donor for the mannose residues is dolichol phosphomannose. The donor for ethanolamine phosphate is the common phospholipid, phosphatidyl ethanolamine, a lipid usually associated with the cytoplasmic leaflet of the lipid bilayer.

A protein to which a glycolipid anchor is to be attached is initially synthesized as a type I transmembrane protein. Such a protein is directed to the endoplasmic reticulum by a cleaved, N-terminal signal sequence and is initially anchored to the membrane by a C-terminal, hydrophobic stop transfer sequence. Attachment of the glycolipid anchor occurs by concerted cleavage of the polypeptide chain and formation of an amide bond with the amino group of ethanolamine. This transamidation reaction requires no additional input of energy. The C-terminal sequences of proteins destined to be processed by the transamidating enzyme are composed largely of non-polar amino acids in order to fulfil their membrane anchor function. Although there is no specific sequence motif associated with glycolipid anchor attachment, the C-terminal amino acid after cleavage and the residue one beyond the cleavage site generally have small side chains. The C-terminal membrane anchors on the precursor forms of these proteins are not followed by hydrophilic cytoplasmic domains.

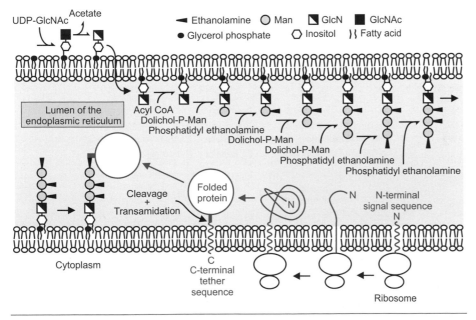

Figure 4.7 Pathway of GPI anchor biosynthesis and attachment to proteins. Anchor synthesis is initiated on the cytoplasmic side of the endoplasmic reticulum membrane and completed on the luminal side. The protein is directed to the lumen by an N-terminal signal sequence that is removed co-translationally to expose a new N-terminal residue (N'). At the end of translation, the protein remains tethered to the membrane by a hydrophobic C-terminal amino acid sequence. Transfer to the glycolipid anchor occurs by proteolytic cleavage of this tether sequence and concerted transfer of the new C-terminal amino acid (C') directly on to the ethanolamine portion of the glycolipid anchor. UDP, uridine diphosphate.

4.8   Proteins attached to glycolipid anchors are localized to the plasma membrane

There is no common theme that relates all the proteins linked to the plasma membrane by GPI anchors. In mammalian cells, a variety of proteins are attached to glycolipid anchors, with no one protein representing a predominant membrane component. What the GPI-anchored proteins do have in common is an almost exclusive localization to the plasma membrane. This distribution can probably be explained by the fact that the absence of a cytoplasmic domain on these proteins makes them invisible to the machinery that mediates endocytosis and other types of traffic between intracellular and plasma membranes.

In spite of the absence of cytoplasmic-facing routing signals on GPI-anchored proteins, they are localized to restricted regions of the plasma membrane, and in most cases, they are enriched in lipid rafts. The structure of the GPI anchor controls movement through the cell, because one of the ethanolamine groups and the inositol-linked fatty acid must be removed before the GPI-anchored protein can exit the Golgi apparatus (Figure 4.8). In addition, the lipid must be remodelled in the Golgi apparatus so that it has two saturated fatty acid chains to allow association with lipid rafts.

In polarized cells, glycolipid-linked proteins are directed almost exclusively to the apical surface. Because this sorting cannot be based on a signal exposed to the cytoplasmic face of the membrane, it has been proposed that the cues for correct targeting are found in the bilayer itself, in the form of the lipid rafts.

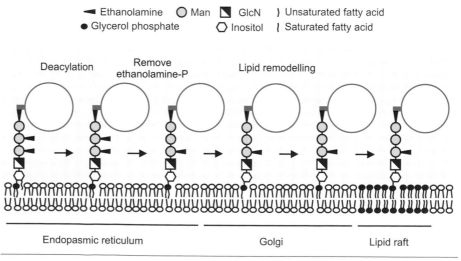

Figure 4.8 Relationship between GPI anchor processing and cellular trafficking. Removal of ethanolamine and acyl groups is required for passage through the secretory system, while remodelling of the fatty acids is needed to allow the anchor to interact with lipid rafts.

The mechanism for directing the glycolipid rafts to vesicles destined for the apical plasma membrane remains unclear, but glycolipid-anchored proteins incorporated into such patches would be carried along to the apical end of the cell.

On mammalian cells, only a modest fraction of the integral membrane proteins in the plasma membrane are linked to glycolipid anchors. In contrast, a very high proportion of the surface coat proteins of trypanosomal parasites are linked in this way. The most abundant surface structures are GPI-linked variant surface glycoproteins. Distinct surface glycoproteins with similar overall structures are expressed from a family of genes. Repeated changes in the expression of these genes allow the parasite to evade the host immune response. The columnar structure of the surface glycoproteins allows them to pack laterally in very dense arrays, effectively shielding the surface of the lipid bilayer from the host. Because the plasma membrane must contain certain essential proteins such as solute transporters, the use of glycolipid anchors provides a way to project the coat proteins from the cell surface while leaving room in the lipid bilayer and the juxtamembrane space for these other proteins (Figure 4.9).

See section 8.3 for more on trypanosome coat proteins.

Variant surface glycoproteins

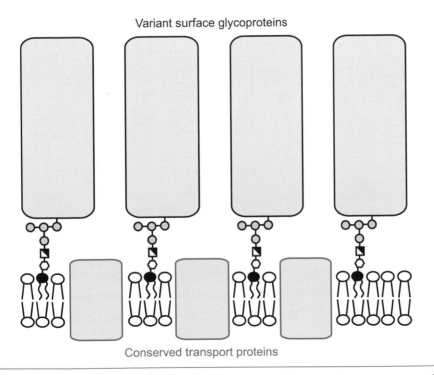

Conserved transport proteins

Figure 4.9 GPI-anchored proteins at the surface of trypanosomal plasma membranes as a means of masking the extracellular portions of transport proteins. Packing of the extended variant surface glycoproteins may prevent access of antibodies to proteins embedded in the lipid bilayer.

LIVERPOOL JOHN MOORES UNIVERSITY
LEARNING SERVICES

BOX 4.2   Glycotherapeutics: *Selective inhibition of glycolipid anchor biosynthesis may provide a way to treat African sleeping sickness*

Glycolipid anchors on variant surface glycoproteins in the plasma membranes of trypanosomes have been extensively studied, and their structures and routes of biosynthesis are now well characterised. Subspecies of *Trypanosoma brucei*, which are transmitted by the tsetse fly, cause up to half a million cases of African sleeping sickness in humans each year as well as the disease nagana in cattle of sub-Saharan Africa.

Two important features of the glycolipid anchors make them appealing targets for drug development aimed at controlling these diseases. First, the anchors are essential for parasite survival in their mammalian hosts. Gene knockout and RNA interference approaches have been used to show that parasites lacking specific enzymes in the glycolipid anchor biosynthesis pathway are not viable in mammals. Second, there are significant differences between the biosynthetic pathways for glycolipid anchors in parasites and mammals. This feature is important, because it suggests that the trypanosomes can be selectively targeted to get around the fact that glycolipid anchors are essential for viability of the mammalian hosts. Comparison of the biosynthetic pathways reveals three major differences. At the beginning of the pathway, the order of removal of the *N*-acetyl group from GlcNAc and acylation of the inositol ring is reversed. Later in the pathway, the order of deacylation and attachment of the ethanolamine phosphate bridge to protein are switched and only the trypanosome anchors undergo fatty acid remodelling (Figure A). These steps may be good targets for development of inhibitors that affect parasite enzymes without affecting mammalian ones.

Figure A

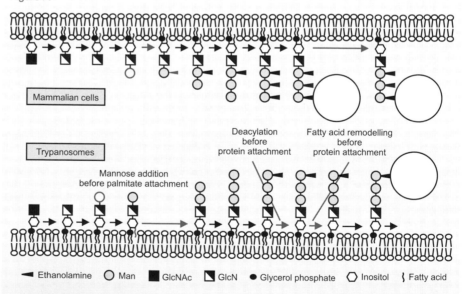

◀ Ethanolamine   ○ Man   ■ GlcNAc   ◣ GlcN   ● Glycerol phosphate   ⬡ Inositol   ⦃ Fatty acid

Some potential antiparasitic compounds that are currently under investigation selectively inhibit the parasite pathway of glycolipid anchor biosynthesis by taking advantage of the fact that the GlcNAc de-*N*-acetylase in the parasites accepts a broader range of substrates than the human enzyme does. Substrate analogues with *N*-carbamyl groups can inhibit the de-*N*-acetylase of both humans and trypanosomes, by acting as suicide substrates that react

Box 4.2   65

covalently with the enzyme active site. However, molecules in which GlcNAc is in β rather than α linkage to inositol, or which contain a methyl group on the inositol ring, are substrates only for the parasite enzyme, so attachment of the *N*-carbamyl group to these substrates makes them selective inhibitors of the parasite enzyme (Figure B). Although these inhibitors work at concentrations as low as 8 nM, they would not be effective *in vivo* because the negatively charged phosphate group makes them membrane-impermeant. To address this issue, an acetoxymethyl group can be attached to the phosphate to neutralize the charge. This group is efficiently removed by cytoplasmic esterases. A protected compound containing a free amino group on glucosamine is able to enter cells and is modified by subsequent enzymes in the pathway, so it can be hoped that it will be possible to achieve killing of trypanosomes using a membrane-permeable version of a covalent inhibitor such as the carbamate derivative.

Figure B

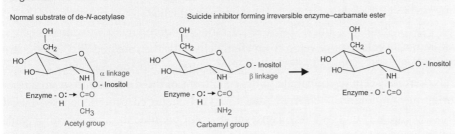

Essay topics

- Discuss the genetic and chemical evidence that the glycolipid anchor biosynthesis pathway is essential for trypanosomal survival in mammalian but not insect hosts.

- Compare the state of development of trypanosome-specific inhibitors of early stages in GPI anchor biosynthesis with the potential for inhibitors of later steps, such as addition of the second and third mannose residues.

- Using the work on trypanosome-specific inhibitors of glycolipid anchor biosynthesis as a model, discuss how a similar strategy might be applied to development of antifungal agents.

Lead references

- Ferguson, M.A.J. (2000). Glycosylphosphatidylinositol biosynthesis validated as a drug target for African sleeping sickness, *Proceedings of the National Academy of Sciences U.S.A.* **97**, 10673–10675.

- Grimme S.J., Colussi P.A., Taron C.H., and Orlean P. (2004). Deficiencies in the essential Smp3 mannosyltransferase block glycosylphosphatidylinositol assembly and lead to defects in growth and cell wall biogenesis in *Candida albicans*, *Microbiology* **150**, 3115–3128.

- Smith, T.K., Crossman, A., Borissow, C.N., Paterson, M.J., Dix, A., Brimacombe, J.S., and Ferguson, M.A.J. (2001). Specificity of the GlcNAc-PI de-*N*-acetylase of GPI biosynthesis and synthesis of parasite-specific suicide substrate inhibitors, *EMBO Journal* **20**, 3322–3332.

- Smith, T.K., Crossman, A., Brimacombe, J.S., and Ferguson, M.A.J. (2004). Chemical validation of GPI biosynthesis as a drug target against African sleeping sickness, *EMBO Journal* **23**, 4701–4708.

- Urbaniak, M.D., Yashunsky, D.V., Crossman, A., Nikolaev, A.V., and Ferguson, M.A.J. (2008). Probing enzymes late in the trypanosomal glycosylphosphatidylinositol biosynthetic pathway with synthetic glycosylphosphatidylinositol analogues, *ACS Chemical Biology* **3**, 625–634.

4.9   The disease paroxysmal nocturnal haemoglobinuria is caused by a glycolipid anchor deficiency

The disease paroxysmal nocturnal haemoglobinuria is associated with the absence of GPI anchors in blood cells owing to a mutation in the *PIG-A* gene, which encodes the GlcNAc-transferase that catalyses the first step in GPI anchor biosynthesis. This is an acquired disease, because the mutation is not present in the germ line but occurs as a somatic mutation in some bone marrow stem cells. The symptom that gives the disease its name results from lysis of red blood cells by the complement system following accidental activation. Such misdirected activation of complement is normally prevented by a series of red blood cell surface proteins, including CD55 (also known as decay accelerating factor) and CD59, which are attached to GPI anchors. If the anchors are not attached, these proteins are absent from the cell surface and a low level of random complement activation occurs. White blood cells that develop from stem cells with the mutated *PIG-A* gene are also deficient in GPI-anchored cell surface proteins, but the deficiency on these cells does not seem to contribute to the symptoms of the disease. Paroxysmal nocturnal haemoglobinuria clearly shows that GPI-anchored proteins on red blood cells have important functions. However, it is not possible to draw any more general conclusions about the functions of GPI anchors from analysis of the symptoms of this disease because glycolipid anchors are normal on cells other than blood cells.

SUMMARY

A good deal more is known about the structure and biosynthesis of membrane-associated glycans than is known about their functions. The finding that membrane glycosylation is not essential for cell viability confirms the notion that many glycans function in an organismal rather than a cellular context. It is also clear that membrane glycoconjugates can mediate and modulate cell–cell interaction, both by affecting the physical properties of cell surfaces and by engaging in specific recognition events. Both these aspects of glycan function are considered in more detail in Chapters 8–10.

KEY REFERENCES

Ferguson, M.A.J. (1999). The structure, biosynthesis and functions of glycosylphosphatidy-linositol anchors, and the contributions of trypanosome research, *Journal of Cell Science* **112**, 2799–2809. This paper provides a detailed overview of GPI-anchored proteins in trypanosomes.

Fujita, M. and Kinoshita, T. (2009). Structural remodeling of GPI anchors during biosynthesis and after attachment to proteins, *FEBS Letters* **584**, 1670–1677. A summary of the biosynthetic pathway for GPI anchors is presented along with evidence for their role in the movement of proteins through cells and into lipid rafts.

Landolt-Marticorena, C. and Reithmeier, R.A.F. (1994). Asparagine-linked oligosaccharides are localised to single extracytosolic segments in multi-span membrane proteins, *Biochemical Journal* **302**, 253–260. This paper presents a survey of glycosylation sites in mammalian polytopic membrane proteins allowing general conclusions to be drawn.

Lingwood, D. and Simons, K. (2010). Lipid rafts as a membrane-organising principle, *Science* **327**, 46–50. This paper provides a physiochemical discussion of what lipid rafts are and what holds them together.

Lopez, P.H.H. and Schnaar, R.L. (2009). Gangliosides in cell recognition and membrane protein regulation, *Current Opinion in Structural Biology* **19**, 549–557. A critical analysis of the main functions of glycolipids is given.

McConnville, M.J. and Ferguson, M.A.J. (1993). The structure, biosynthesis and function of glycosylated phosphatidyl inositols in the parasitic protozoa and higher eukaryotes, *Biochemical Journal* **294**, 305–324. This paper provides a detailed overview of GPI-anchored proteins.

Merrill, A.H., Jr., Wang, M.D., Park, M., and Sullards, M.C. (2007). (Glyco)sphingolipidology: an amazing challenge and opportunity for systems biology, *Trends in Biochemical Sciences* **32**, 457–468. An overview is given of the complexity of glycosphingolipid structure and biosynthesis.

Schwarzmann, G. and Sandhoff, K. (1990). Metabolism and intracellular transport of glyco-sphingolipids, *Biochemistry* **29**, 10865–10871. The biochemistry and cell biology of glyco-sphingolipid metabolism are reviewed.

Tomita, M. (1999). Biochemical background of paroxysmal nocturnal hemoglobinuria, *Biochimica et Biophysica Acta* **1455**, 269–286. The genetic and biochemical basis for the symptoms of this glycolipid anchor deficiency disease are reviewed in detail.

Winchester, B., Vellodi, A. and Young, E. (2000). The molecular basis of lysosomal storage diseases and their treatment, *Biochemical Society Transactions* **28**, 150–154. A concise review is provided of glycolipid storage disorders and potential treatments.

Yamashita, T., Wada, R., Sasaki, T., Deng, C., Bierfreund, U., Sandhoff, K., and Proia, P.L. (1999). A vital role for glycosphingolipid synthesis during development and differentiation, *Proceedings of the National Academy of Sciences U.S.A.* **96**, 9142–9147. This paper describes the phenotype of mice in which the gene for glucosylceramide synthase has been knocked out, preventing synthesis of glycosphingolipids.

QUESTIONS

4.1   What are the similarities and differences between the pathways for biosynthesis of glycolipids and those for O-linked glycans and N-linked glycans?

4.2   Go to the KEGG Pathway Database at http://www.genome.jp/kegg/pathway.html and study the pathways for glycosphingolipid metabolism. In particular, examine gan-glioside biosynthesis and note the many different structures built up from lactosyl-ceramide. You can get further information on some of the enzymes by clicking on their numbers. How many different sialyltransferases are needed to synthesize the ganglioside series?

4.3   Compare the structures and functions of glycans attached to glycolipids with those of glycoproteins.

4.4   Discuss the evidence that gangliosides concentrated into plasma membrane micro-domains modulate signalling pathways.

Lead reference: Allende, M.L. and Proia, R.L. (2002). Lubricating cell signalling pathways with gangliosides, *Current Opinion in Structural Biology* **12**, 587–592.

Enzymology and cell biology of glycosylation

5

LEARNING OBJECTIVES

By the end of this chapter, students should understand:

1 How biochemistry, novel cloning strategies, and genomics have been combined to characterize proteins required for glycosylation

2 Topological and mechanistic features of the glycan biosynthesis pathway

3 The organization of the glycan biosynthesis pathway in cells and how it is established

4 The importance of mutant cell lines and knockout mice in establishing fundamental ideas about the biological roles of glycosylation

The ways in which cells generate glycoproteins and glycolipids provide important clues about how information can be encoded in these structures. Much recent progress in understanding glycoconjugate biosynthesis has resulted from parallel developments in molecular and cell biology. This chapter presents a summary of some key steps in identifying the enzymes that form glycan biosynthesis pathways, followed by a discussion of how these steps are integrated in the secretory pathways of cells. This knowledge also facilitates manipulation of the structures for experimental and therapeutic purposes.

5.1 Only a few eukaryotic glycosyltransferases are sufficiently abundant to be isolated biochemically

Many glycosyltransferases are expressed in cells throughout the body, but few are expressed at high levels in any specific tissues. An exception, which provided an early breakthrough in the characterization of these enzymes, is galactosyltransferase. This enzyme is expressed at high levels in mammary tissue, where it forms one of the subunits of the enzyme lactose synthetase, which utilizes UDP-Gal as donor and glucose as acceptor for the synthesis of lactose (Figure 5.1). Lactose synthetase can be isolated from milk in soluble form using conventional chromatographic methods and consists of two subunits, both of which are required for lactose synthesis activity. However, the dissociated A subunit is able to act as a galactosyltransferase, utilizing glycoproteins bearing exposed GlcNAc as acceptors and forming terminal Galβ1-4GlcNAc linkages. The B subunit, which is also found free in milk as the protein α-lactalbumin, acts as a specifier protein that changes the acceptor specificity to glucose. Thus lactose synthetase purified from milk, and dissociated into A and B subunits, is a readily available source of galactosyltransferase protein. Clones for the full-length protein, isolated from a mammary gland expression library using antibodies raised against the purified A subunit, encode a type II transmembrane protein. This full-length galactosyltransferase is found in the Golgi apparatus. The soluble A subunit of lactose synthetase in milk represents the luminal domain of the protein released by proteolysis.

Sialyltransferases are the only other type of glycosyltransferase that can be purified directly from mammalian tissues. Submaxillary glands are a rich source of enzyme, because of the abundance of sialylated O-linked glycans on the mucins made in this tissue. These enzymes do not work on low molecular weight acceptors, but their activity can be assayed using mucins that have been desialylated by acid treatment to expose terminal galactose residues (Figure 5.2). It is necessary to use detergent to solubilize the intracellular forms of the enzymes and a key step in their purification is use of affinity chromatography on a substrate analogue. Immobilized CDP serves to mimic part of the CMP-NeuAc donor, in contrast to resins based on acceptor mimics, which are not effective. In addition to α2-3 sialyltransferase purified from submaxillary glands in this way, two other sialyltransferases can be purified using similar strategies, resulting in sufficient protein for partial sequence analysis and cloning. All

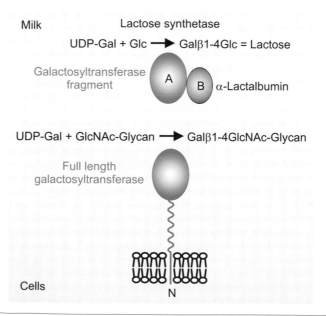

Figure 5.1 Organization of milk lactose synthetase and cellular galactosyltransferase. The catalytic domain of the Golgi galactosyltransferase is released in soluble form by proteolysis and takes on lactose synthetase activity when complexed with the modifier protein α-lactalbumin.

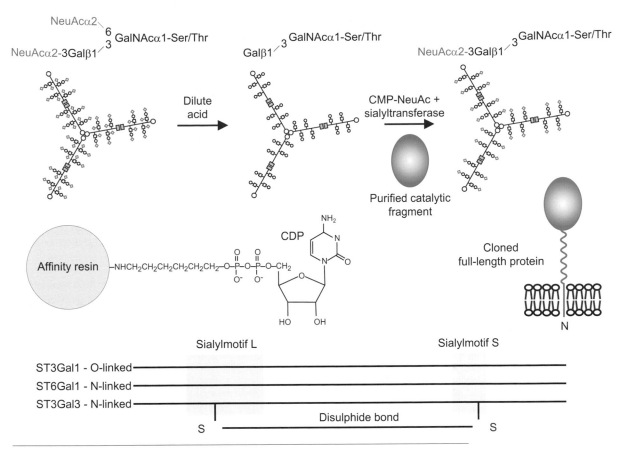

Figure 5.2 Tools used in the isolation and cloning of sialyltransferases. Acceptor substrate useful in assaying the enzyme can be created by mild acid hydrolysis of mucin to release sialic acid residues, but affinity purification requires use of a donor substrate analogue. Conserved residues forming a sialylmotif, observed by comparison of sequences of three sialyltransferases, can be used in identification of additional members of the family.

these enzymes are type II transmembrane proteins, but they share no more that 15% overall sequence identity, which is little more than would be expected for unrelated proteins. However, two partially conserved regions, designated the L and S sialylmo-tifs, can be used to design primers for amplification of other transferase cDNAs with the polymerase chain reaction. Expression of these proteins and testing on a panel of potential substrates, including glycoproteins and glycolipids, demonstrates different acceptor specificities in the family of proteins defined by the sialylmotifs.

5.2 Novel molecular biology methods are needed to clone additional glycosyltransferases

Biochemical purification of galactosyltransferase and sialyltransferases, as well as one GlcNAc transferase, provides insight into the enzymology of these enzymes

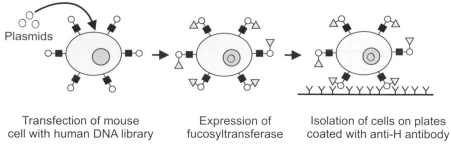

Transfection of mouse
cell with human DNA library

Expression of
fucosyltransferase

Isolation of cells on plates
coated with anti-H antibody

Figure 5.3 Isolation of the gene for the fucosyltransferase needed to generate the H substance in the ABO blood group system. Transfection of a panel of human genes into mouse cells, which do not make the H glycan, is followed by selection of cells able to adhere to antibody that recognizes this glycan.

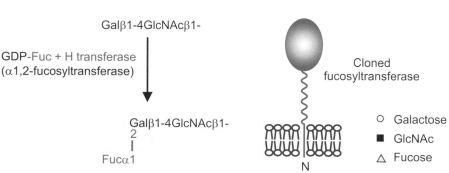

➲ See **Section 2.7** for more on Bombay blood type.

and enables development of antibodies used to study their intracellular localization, though there are limitations to the purify-and-clone strategies because of the low abundance of most glycosyltransferases. Novel gene transfer cloning strategies in eukaryotic cells provide an important alternative route to glycosyltransferase identification (Figure 5.3). In this approach cells that do not express a specific glycosyltransferase, but which are able to make glycoprotein or glycolipid acceptors, are transfected with clones from a genomic library created in an expression vector and are then screened for the presence of the product glycan made by the enzyme. For cloning of the human blood group H transferase, mouse cells are a suitable recipient because mice do not make the blood group H oligosaccharide. Enzyme activity is detected by screening with an antibody to the H substance, isolated from individuals with the rare Bombay blood type who also do not make the H oligosaccharide and hence make antibodies to it. Following multiple rounds of re-screening, individual clones that confer expression of H antigen can be isolated and the expected type of fucosyltransferase activity is found in lysates of the transfected cells.

Similar strategies employing glycan-specific lectins, as well as antibodies as detection reagents, have led to cloning of other mammalian glycosyltransferases. Characterization of other proteins involved in glycan biosynthesis, including processing glycosidases, requires similar combinations of specialized biochemical and cloning techniques. Identification of nucleotide sugar transporters, for example, is based on development of lipid reconstitution methods to demonstrate their ability to move sugar donors across membranes.

5.3   Structurally diverse glycosyltransferases share key common features

In spite of their overall sequence diversity, all known eukaryotic glycosyltransferases share a common organization as type II transmembrane proteins. No functional eukaryotic glycosyltransferases have been expressed in bacteria, but study of enzymes expressed in eukaryotic expression systems and analysis of bacterial homologues have provided structural insights into how these enzymes work. As well as having a common arrangement in the membrane, mammalian glycosyltransferases fall into two general structural categories and in each category the fold of the catalytic domain is similar to folds observed for bacterial enzymes that transfer sugars onto oligosaccharides as well as other types of acceptors such as antibiotics.

Many glycosyltransferases also share features in their catalytic centres. One common characteristic is sequential binding of substrates, in which the sugar nucleotide donor binds first, leading to ordering of a disordered loop, which in turn creates the binding site for the acceptor substrate (Figure 5.4). This arrangement explains why affinity purification of glycosyltransferases by chromatography on donor analogues is more successful than use of acceptors. Another common feature is a pair of acidic residues at characteristic spacing to form DXD motifs that ligate metal ions which in turn ligate to the nucleotide sugar donor.

Some of the most extensively studied glycosyltransferases are the enzymes that attach the terminal GalNAc and galactose residues to ABO blood group oligosaccharides.

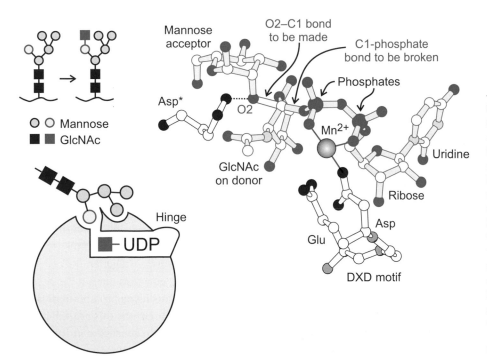

Figure 5.4 Structure of GlcNAc-transferase I. The enzyme adds a GlcNAc residue to a partially trimmed high mannose oligosaccharide. The UDP-GlcNAc donor substrate enters the binding site first and a conformational change at the hinge produces a binding site for the acceptor substrate on top of the donor. The DXD motif, which mediates binding of the nucleotide sugar donor, is characteristic of glycosyltransferases. Asp* catalyses the transfer reaction. [Based on entry 1FOA in the Protein Databank.]

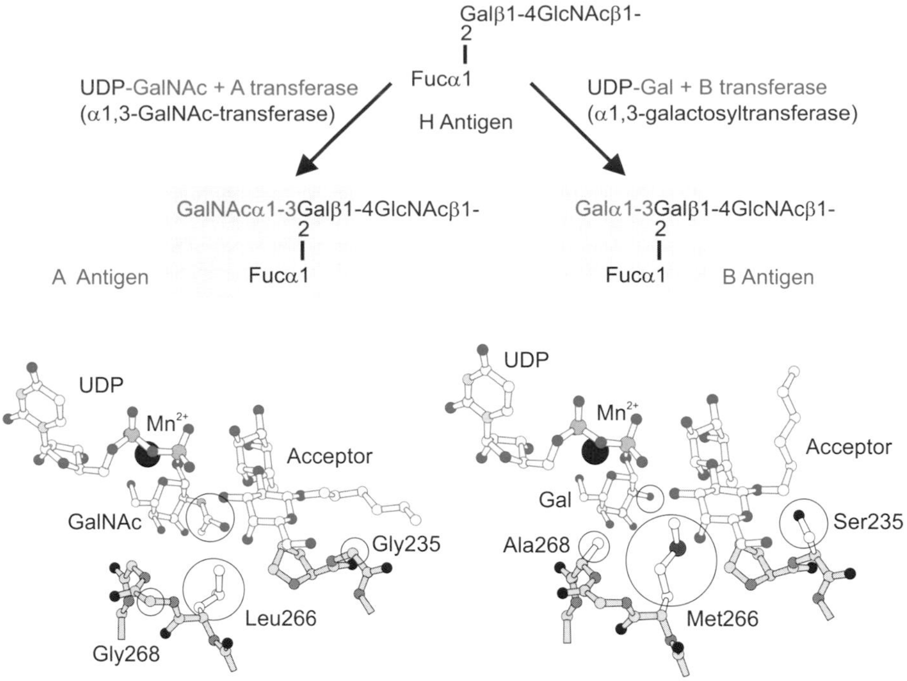

Figure 5.5 Differences between the A and B transferases. Amino acid substitutions at positions 266 and 268 change the specificity of the A enzyme from using UDP-GalNAc as substrate to using UDP-Gal, possibly because the larger side chains at these positions clash with the large substituent on the 2 position of GalNAc. [Based on entries 1LZI and 1LZJ in the Protein Databank and Patenaude, S.I., Seto, N.O.L., Borisova, S.N., Szpacenko, A., Marcus, S.L., Palcic, M.M., and Evans, S.V. (2002). The structural basis for specificity in human ABO(H) blood group synthesis, *Nature Structure Biology* **9**, 685–690.]

⊙ See Section 2.7 for more on ABO blood groups.

There are four amino acid differences between the A and B transferases. Changing all four residues can change the GalNAc specificity of the A enzyme into the galactose specificity of the B enzyme. In fact, making just the changes Leu266 to Met and Gly268 to Ala in the donor binding site is sufficient to create the swapped specificity, probably because the longer side chains prevent binding of the bulkier substrate (Figure 5.5). However, making the reverse changes in the B enzyme does not create an effective A enzyme, and other combinations of mutations create dual specificity enzymes.

5.4 Genomics can be used to define the repertoire of glycosyltransferases

⊙ See Section 6.5 for more on genomics of glycosyltransferases.

⊙ See Section 6.7 for more on glycan arrays.

Sequence signatures, such as the sialylmotifs, common to specific types of glycosyltransferases provide a way to screen the human genome sequence to identify additional members of each family. These studies define the overall complement of glycosylating enzymes present in humans and other organisms. In general, sequences can usually be used to predict the donors used by specific enzymes but not the acceptors. Experimental approaches are needed to define acceptors using proteins expressed from genes predicted to encode glycosyltransferases. Ways of doing this are still being developed, but they can include testing for sugar transfers onto substrates individually or in panels on glycan arrays.

5.5   Special transporters are required to provide access of donor substrates to glycosyltransferases

In the initial stages of N-linked glycan and glycolipid biosynthesis, access of donor substrates to glycosyltransferases in the endoplasmic reticulum is provided by flip-pases that expose the headgroups of dolichol-linked sugars to the luminal compartment. The molecular identity of these transporters remains one of the major unsolved mysteries in defining the pathway for the biosynthesis of glycoproteins and glycolipids. Glycosyltransferases acting at later stages in these processes and in O-linked glycosylation, which occur in the Golgi apparatus, use nucleotide sugars as substrates. Because these donors are synthesized in the cytoplasm, a special set of transporters is required to deliver them to the active sites of the enzymes, which are located on the luminal side of the membrane.

The transporters isolated in vesicles from the Golgi apparatus are antiporters that exchange the donor nucleotide sugars for the sugar-free nucleotides released by the glycosyltransferases (Figure 5.6). As with glycosyltransferases, an important tool in the study of the transporters is the use of gene transfer to correct defects in cells lacking specific transporters. For example, the gene for the mouse transporter for UDP-GlcNAc was identified by transfecting yeast cells that lack this activity. Like the glycosyltransferases, sugar nucleotide transporters share common structural features, including the presence of multiple membrane-spanning sequences, which provide a means of identifying additional transporters by inspection of genome sequences.

⊙ See **Section 2.3** for more on the topology of glycosylation.

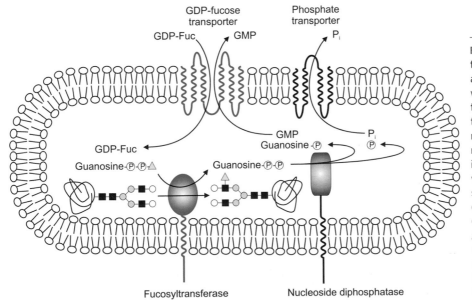

Figure 5.6 Nucleotide sugar transporters in the Golgi apparatus. The transporters, which are exchange proteins that swap nucleotide sugars for the sugar-free monophosphorylated nucleotides, are complex integral membrane proteins with multiple membrane-spanning segments. In the case of fucosyltransferases and other glycosyltransferases that generate diphosphorylated nucleotide products, one phosphate must be removed to allow transport back into the cytoplasm.

BOX 5.1   Glycobiology of disease: *Transporter deficiency can result in aberrant glycosylation*

The transporter that translocates GDP-fucose the donor sugar for fucosyltransferases into the Golgi apparatus was discovered by analysis of cells from patients with a very rare clinical syndrome, leucocyte adhesion deficiency type II (LAD-II). Patients with LAD-II have mental retardation, abnormalities of the skeleton leading to short stature, and immunodeficiency resulting in susceptibility to bacterial infection. Analysis of their glycoproteins reveals that fucose-containing glycans are absent and they have the unusual 'Bombay' blood type instead of the usual A, B, or O because they are unable to synthesize the starting H-structure (see Chapter 2).

The susceptibility to infection is due to the leucocyte adhesion deficiency. LAD II patients have high levels of circulating neutrophils and other types of white blood cell, because these cells are unable to migrate out of the blood stream into the tissues to sites of infection. Migration of neutrophils out of the circulation first requires adhesion to the endothelial cells lining the blood vessels and this interaction is mediated by a group of glycan-binding receptors, the selectins (see Chapter 9), which bind to fucosylated glycans that are absent in LAD II patients.

The defective gene in LAD II was identified by complementation studies in which transfection of a cDNA into a LAD II patient's fibroblasts was found to correct the fucosylation defect. The protein identified as the human GDP-fucose transporter has ten membrane-spanning regions with both the amino and carboxy termini in the cytoplasm. It transports GDP-fucose from the cytoplasm, where it is synthesized, into the lumen of the Golgi apparatus in exchange for GMP. Five different mutations in the GDP-fucose transporter have been found in seven children diagnosed with LAD II . These mutations either result in single amino acid substitutions that make the transporters non-functional, or cause premature stop codons that result in truncated transporters that are mis-localized to the endoplasmic reticulum as well as being inactive. A generalized defect in fucosylation of glycans results, because none of the fucosyltransferases that act on glycoproteins or glycolipids in the Golgi apparatus have access to their donor substrates. The lack of neutrophil migration and increased bacterial infections in LAD II can be fully explained by the absence of fucosylated ligands for the selectins. In contrast, the specific molecular mechanisms underlying the developmental defects in LAD II are unknown, but these defects may point to roles for terminal fucose residues on glycans in cell–cell interactions during development. Because the underlying defect in these individuals is a problem in protein glycosylation, the alternative designation congenital disorder of glycosylation type IIc has been suggested (Chapter 14).

Treatment of one LAD II patient with high doses of oral fucose was successful in curing the immunodeficiency and the child also had some reduction in severity of neurological defects. Fucosylation of cell surface glycans, including the selectin ligands, was restored, allowing migration of neutrophils to sites of infection. However, fucose treatment in two other patients had no effect on glycosylation. In a fourth patient, fucosylated glycans were produced as a result of oral fucose therapy, but unfortunately this patient developed an autoimmune response due to synthesis of the H-antigen which was recognized by antibodies present in the patient's blood. High doses of oral fucose increase the amount of GDP-fucose synthesized in the cytoplasm, but it is not known how the GDP-fucose was transported into the Golgi apparatus in the two patients who responded to treatment. There must be an alternative way of transporting GDP-fucose. A

Box 5.1   77

second GDP-fucose transporter that localizes to the endoplasmic reticulum has been identified in *Drosophila* but a mammalian homologue has not yet been found.

Essay topic

- Discuss how analysis of LAD II has increased our understanding of the molecular mechanisms underlying fucosylation of glycans and provided evidence for the function of fucose-containing glycans in cellular recognition events.

Lead references

- Etzioni, A., Frydman, M., Pollack, S., Avidor, I., Phillips, M.L., Paulson, J.C., and Gershoni-Baruch, R. (1992). Recurrent severe infections caused by a novel leukocyte adhesion deficiency, *New England Journal of Medicine* **327**, 1789–1792.

- Helmus, Y., Denecke, J., Yakubenia, S., Robinson, P., Luhn, K., Watson, D.L., McGrogan, P.J., Vestweber, D., Marquardt, T., and Wild, M.K. (2006). Leukocyte adhesion deficiency II patients with a dual defect of the GDP-fucose transporter, *Blood* **107**, 3959–3966.

- Lübke, T., Marquardt, T., Etzioni, A., Hartmann, E., von Figura, K., and Körner, C. (2001). Complementation cloning identifies CDG-IIc, a new type of congenital disorder of glycosylation, as a GDP-fucose transporter deficiency, *Nature Genetics* **28**, 73–76.

- Lühn, K., Wild, M.K., Eckhardt, M., Gerardy-Schahn, R., and Vestweber, D. (2001). The gene defective in leukocyte adhesion deficiency II encodes a putative GDP-fucose transporter, *Nature Genetics* **28**, 69–72.

5.6   Complex cellular machinery is required for spatial and temporal organization of glycan biosynthesis pathways

The order of addition of sugars to growing glycan chains is determined both by the substrate specificities of the enzymes and their sequential access to substrates as they move through the secretory pathway. Thus, the organization of this machinery within eukaryotic cells plays an important role in determining which glycans are made. Sites at which different steps in biosynthesis occur can be identified by staining cells with lectins that recognize specific intermediate glycans and with antibodies to the biosynthetic enzymes. For example, the GalNAc-transferases that initiate mucin-type O-glycans stain the Golgi apparatus, which is also the site at which a lectin specific for GalNAc linked to serine or threonine residue binds (Figure 5.7). Within the Golgi apparatus, the order of action of enzymes in the biosynthetic pathways for N- and O-linked sugars is reflected in their position in the Golgi stack: mannosidase I is found in the *cis* Golgi, while the GlcNAc-transferases that establish the branching structures are found in the medial Golgi and enzymes that add terminal galactose, and sialic acid residues are found in the *trans* Golgi.

Because of the key role of the Golgi apparatus in glycoconjugate biosynthesis, our understanding of this organelle has developed in parallel with study of the enzymes of glycan biosynthesis. Localization signals in different glycosyltransferases are

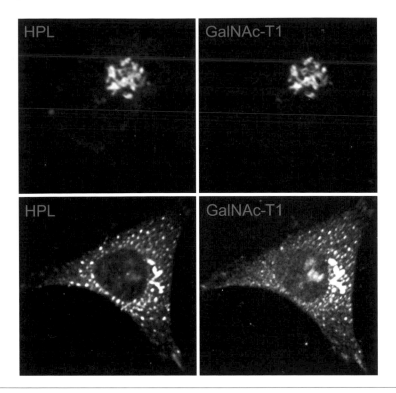

Figure 5.7 Localization of a glycosyltransferase and its product in cells by fluorescence microscopy. Co-localization of GalNAc linked to serine and threonine residues, stained with the lectin HPL from the snail *Helix pomatia*, and one of the GalNAc transferases that initiate mucin-type O-glycans stained with antibodies. In the upper panels, both the enzyme and its product are found in the Golgi apparatus in untreated HeLa cells. In the lower panels, both appear in the endoplasmic reticulum as well following treatment with epidermal growth factor. [Reproduced from Gill, D.J., Chia, J., Senewiratne, J., and Bard, F. (2010) Regulation of O-glycosylation through Golgi-to-ER relocation of initiation enzymes, *Journal of Cell Biology* **189**, 843–858, with permission.]

found in the cytoplasmic tails, the transmembrane domains, the stalk regions, and the catalytic domains, but there are multiple explanations for how these signals work (Figure 5.8). One possibility is that the length of the transmembrane domain in each glycosyltransferase determines its optimal position as a result of a gradient of increasing bilayer thickness from the endoplasmic reticulum through the Golgi apparatus to the plasma membrane. If the Golgi is viewed as a stationary organelle through which substrates pass, a key feature of the enzymes may be their exclusion from vesicles that transport the substrates from one Golgi layer to the next. It is possible that such exclusion results from formation of large complexes of enzymes. Alternatively, in the cisternal progression model of the Golgi apparatus, glycosyltransferases must be sorted into transport vesicles for retrieval and retrograde transport, possibly by binding to adapter molecules that interact with the cytoplasmic tails of the enzymes. Changes in the localization of enzymes may also alter the pattern of glycosylation. For example,

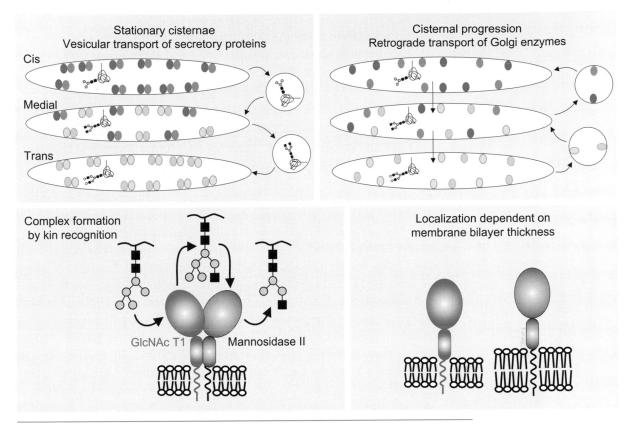

Figure 5.8 Mechanisms to control localization of glycosyltransferases at specific sites in the secretory pathway. In a stationary Golgi apparatus, movement of glycosyltransferases may be restricted by formation of complexes. Alternatively, if movement of substrates results from progression of the cisternae, retrograde transport of glycosyltransferases could be mediated by adapters that recognize retention signals in the cytoplasmic domains. Matching the length of the transmembrane domain with bilayer thickness, which increases along the pathway, is one proposed mechanism for determining the sites of retention of glycosyltransferases.

stimulation of cells with growth factors can lead to movement of GalNAc-transferases from the Golgi apparatus to the endoplasmic reticulum (Figure 5.7). Similar changes in tumour cells may be associated with altered glycosylation in these cells.

In addition to localization of enzymes in specific portions of the secretory pathway, local organization of glycosyltransferases in complexes may facilitate specific sequences of enzymatic steps in glycoprotein maturation. Two enzymes that associate with each other are GlcNAc-transferase I, which adds the first GlcNAc to an N-linked glycan as it is processed from high mannose to complex forms, and mannosidase II, which catalyses the subsequent removal of two mannose residues. The neck domains of these two enzymes associate on the luminal side of the membrane in a process sometimes referred to as **kin recognition**. Other enzymes that interact with each other include the galactosyl and GlcNAc transferases that act sequentially in

⊜ See **Section 13.7** for more on glycosylation in cancer.

formation of polylactosamine chains as well as several of the enzymes involved in glycosaminoglycan chain biosynthesis and modification. These associations may ensure that intermediate products are not released and swept onward before both enzymatic steps have occurred, and they may also facilitate processive activity of the enzymes, allowing them to catalyse multiple rounds of additions and modifications.

5.7   Mutant cell lines serve as tools for studying glycosylation and illustrate the importance of N-linked glycosylation

Chinese hamster ovary cells with altered glycosylation machinery have been generated by chemical and radiation mutagenesis. These hamster cells are favoured for such studies because they are diploid, genetically stable, and have a low chromosome number (2n = 22). The key step in creation of glycosylation mutants is selection using toxic plant lectins, bacterial toxins, or antibodies to carbohydrates. Selective agents that have been used include phytohaemagglutinin (a lectin from kidney bean that binds to galactose-terminated bi-antennary complex glycans) and ricin (a lectin from castor beans that binds to any glycans with terminal β1-4-linked galactose residues).

Selection with these toxins results in mutant cell lines defective in several different glycosylation steps leading to galactose-terminated glycans. In some cases, the changes specifically affect the galactose-bearing glycans, while in other cases, they affect broader classes of glycans (Figure 5.9). Such strategies have been used to generate lectin-resistant cell lines with mutations in glycosyltransferase genes, processing glycosidases, nucleotide sugar transport, and nucleotide sugar biosynthesis. For

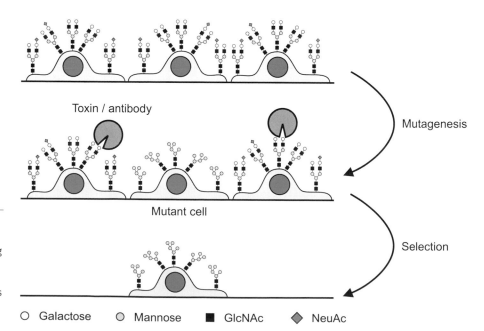

Figure 5.9 Selection of Chinese hamster ovary cells with aberrant glycosylation. Following mutagenesis, cells unable to attach galactose to glycans can be selected using a toxin such as ricin, which targets terminal galactose residues to kill cells.

example, Lec1 cells express exclusively high mannose N-linked glycans because the *GlcNAc-T1* gene has been mutated, and in Lec2 cells, all glycoprotein and glycolipid glycans lack sialic acid because the CMP-sialic acid transporter is inactive.

Glycosylation-deficient cell lines have helped to define the pathways of glycan biosynthesis and they can be used as recipients for gene transfer cloning experiments. They also have been very useful tools in production of glycoproteins with modified glycosylation. Such glycosylation engineering has become particularly important in the area of therapeutic glycoprotein production. Because the structure of glycans on glycoproteins can often determine their targeting and turnover within organisms, generating glycoproteins with particular types of glycosylation can be critical for delivering them to the correct cell types or for ensuring that they have an appropriately long half-life in circulation.

Mutant cell lines have also been critical in defining the functions of different types of glycosylation in cells. For example, the viability of Lec1 cells, on which all N-linked glycans are exclusively the high mannose type, provides direct evidence that complex-type glycans are not required for cell viability. However, N-linked glycosylation cannot be dispensed with altogether. Mutant cells that lack all N-linked glycosylation can be created by breading mice heterozygous for a knockout of GlcNAc-1-phosphotransferase, which is needed to make the UDP-GlcNAc donor used in synthesis of the dolichol-linked precursor for N-linked glycans, by crossing mice that are heterozygous for this mutation. The fact that the cells from homozygous embryos are not viable demonstrates that N-glycosylation is required for fundamental cell functions.

⊜ See **Section 13.5** for more on glycosylation engineering.

⊜ See **Sections 10.7–10.9** for more on glycoprotein targeting.

5.8 Knockout mice provide much of the evidence for the roles of glycosylation in the biology of mammals

Important conclusions about the roles of glycosylation in organisms have been based on extension of the mutagenesis strategy to knockout mice (Figure 5.10). A key finding

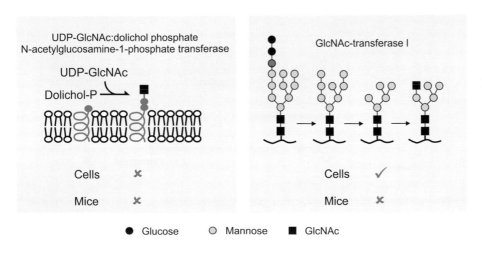

Figure 5.10 Comparison of the effects of eliminating classes of N-linked glycans in cells and organisms. Experiments in cells show that N-linked glycans are required for cell viability, but processing to complex structures is not necessary. In contrast, complex glycans are required for production of viable mice.

is that mice lacking GlcNAc-transferase I die at embryonic day 8.5. Thus, in contrast to the situation in cells, complex glycans are required for organismal development.

Similar types of studies reveal that glycolipids, glycolipid anchors, mucin-type O-glycans, and proteoglycans can all be completely deleted in cells but not in organisms. Knockout of the ceramide glucosyltransferase gene, preventing formation of glucosphingolipids, causes lethality at around the time of gastrulation and mice lacking glycosyl phosphatidylinositol (GPI) anchors on proteins resulting from knocking out the *PIG-A* gene die early in embryogenesis. Because some maternally directed enzyme activity may be present in the egg, allowing synthesis of sufficient glycans for early steps in embryogenesis, the requirements for N-linked glycans may start even earlier than is suggested by the stage at which development of the embryo ceases. The presence of multiple genes for attachment of GalNAc to serine and threonine residues to initiate mucin-type O-glycan synthesis makes it very difficult to undertake similar experiments eliminating this class of glycans altogether. However, elimination of the galactosyltransferase that modifies this GalNAc is lethal. Although it seems likely that almost every class of glycoconjugate is essential

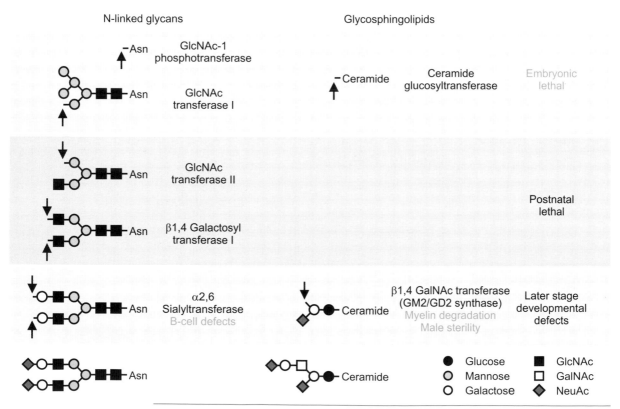

Figure 5.11 Effects of mutations at different steps in glycoconjugate biosynthesis. Mutations in the pathways for N-linked glycan and glycosphingolipid biosynthesis illustrate that elimination of entire classes of glycoconjugates is often lethal in embryonic development, while disruption of terminal modifications tends to affect the behaviour of selected groups of cells in otherwise viable mice.

⊙ See **Section 4.9** for more on GPI anchors.

for steps in early embryonic development, in no case is it known exactly why the glycans are essential.

In contrast to the early lethality associated with eliminating entire classes of glycans, few of the enzymes that add terminal elaborations to these structures are essential for mouse viability. A combination of studies in cells and knockout mice reveals that the phenotypes associated with the absence of specific classes of glycans become progressively less severe as the set of glycans affected becomes smaller (Figure 5.11). Thus, elimination of all N-linked glycans is fatal in cells, while elimination of all complex N-linked glycans is lethal in organisms but not cells. Generation of only partially processed N-linked glycans results in organisms that survive for a limited time after birth, and elimination of specific N-linked glycans terminating in 2-6 linked sialic acid affects primarily a single cell type, B cells. Similarly, knockout of just the glycolipids GM2 and GD2 affects only a subset of cells, causing myelin degradation and problems with sperm development. For many of the terminal sugars, multiple glycosyltransferases can synthesize identical or similar linkages, which may lead to compensation for missing enzymes. However, it should also be emphasized that there are tissue- and stage-specific defects in many of these knockout mice, indicating critical roles for particular glycans in more specialized stages of development. Although the roles of many of the essential glycans are not known, some are described in other chapters. In general, it may prove easier to work out the molecular basis for function of glycans in individual tissues at later stages in development, because biochemical analysis will be more feasible.

SUMMARY

Our knowledge of the complement of glycosyltransferases and sugar nucleotide transporters, as well as other enzymes required for glycan modification in eukaryotic cells, is derived from a very sophisticated set of protein chemistry and molecular and cell biology tools that only became available in the past 25 years. In spite of much progress, the repertoire of enzymes has still not been completely described. Increases in our understanding of the spatial and temporal organization of glycosylation in cells have similarly been tied closely to progress in understanding the organization and function of the Golgi apparatus. Modification of the pathways of glycoconjugate biosynthesis has provided key evidence for the essential roles of glycosylation in cells and organisms.

KEY REFERENCES

Caffaro, C.E. and Hirschberg, C.B. (2006). Nucleotide sugar transporters of the Golgi apparatus: from basic science to diseases, *Accounts of Chemical Research* **39**, 805–812. A summary of both the biochemical characterization and cloning of nucleotide sugar transporters.

Colley, K.J. (1997). Golgi localization of glycosyltransferases: more questions than answers, *Glycobiology* **7**, 1–13. A critical assessment of evidence for different models of how glycosyltransferases become localized to appropriate regions in the Golgi apparatus.

De Graffenried, C.L. and Bertozzi, C.R. (2004). The roles of enzyme localization and complex formation in glycan assembly within the Golgi apparatus, *Current Opinion in Cell Biology* **16**, 356–363. Some updated views of evidence for different theories about how resident and nonresident proteins are sorted in the Golgi apparatus.

Ernst, L.K., Rajan, V.P., Larson, R.D., Ruff, M.M., and Lowe, J.B. (1989). Stable expression of blood group H determinants and GDP-L-fucose:β-D-galactoside 2-α-L-fucosyltransferase in mouse cells after transfection with human DNA, *Journal of Biological Chemistry* **264**, 3436–3447. This article describes a key step in the expression cloning of a glycosyltransferase.

Gordon, R.D., Sivarajah, S., Satkunarajah, M., Ma, D., Tarling, C.A., Vizitiua, D., Withers, S.G., and Rini, J.M. (2006). X-ray crystal structures of rabbit *N*-acetylglycosaminyltransferase I (GnT I) in complex with donor substrate analogues, *Journal of Molecular Biology* **360**, 67–79. An example of a glycosyltransferase that contains a DXD motif, showing the conformational change associated with the binding of the donor substrate.

Kikuchi N. and Narimatsu H. (2006). Bioinformatics for comprehensive finding and analysis of glycosyltransferases, *Biochimica et Biophysica Acta* **1760**, 578–583. A discussion of how sequence motifs can be used to identify glycosyltransferases in the human genome.

Lowe, J.B. and Marth, J.D. (2003). A genetic approach to mammalian glycan function, *Annual Review of Biochemistry* **72**, 643–691. Most of the knockout mice created in glycan biosynthesis pathways are summarized in a well-organized and critical review.

Patenaude, S.I., Seto, N.O.L., Borisova, S.N., Szpacenko, A., Marcus, S.L., Palcic, M.M., and Evans, S.V. (2002). The structural basis for specificity in human ABO(H) blood group synthesis, *Nature Structure Biology* **9**, 685–690. Comparison of the binding sites in the A and B transferases reveals the minor differences needed to change donor substrate specificity.

Ramakrishnan, B. and Qasba, P.K. (2001). Crystal structure of lactose synthase reveals a large conformational change in its catalytic component, the β1-4-galactosyltransferase-I, *Journal of Molecular Biology* **310**, 205–218. A description of the relationship between lactose synthetase and galactosyltransferase.

Rao, F.V., Rich, J.R., Rakić. B., Buddai, S., Schwartz, M.F., Johnson, K., Bowe, C., Wakarchuk, W.W., DeFrees, S., Withers, S.G., and Strynadka, N.C.J. (2009). Structural insights into mammalian sialyltransferases, *Nature Structural and Molecular Biology* **16**, 1186–1188. Structure of a mammalian sialyltransferase illustrating the role of the conserved sialylmotif.

Stanley, P. and Ioffe, E. (1995), Glycosyltransferase mutants: key new insights in glycobiology, *FASEB Journal* **9**, 1436–1444. A summary of some of the insights that have come from analysis of glycosylation mutants, particularly those generated in Chinese hamster ovary cells.

QUESTIONS

5.1   Discuss why a molecular understanding of the glycosylation machinery in cells only developed in the last part of the 20th century.

5.2   The enzyme for production of UDP-xylose, needed for the initial steps of proteoglycan biosynthesis, is found in the lumen of the Golgi, so there would seem to be no need for a UDP-xylose transporter. Discuss the evidence that there is such a transporter and why it might be needed.

Reference: Bakker, H., Oka, T., Ashikov, A., Yadav, A., Berger, M., Rana, N.A., Bai, X., Jigami, Y., Haltiwanger, R.S., Esko, J.D., and Gerardy-Schahn, R. (2009). Functional UDP-xylose transport across the endoplasmic reticulum/Golgi membrane in a Chinese hamster ovary cell mutant defective in UDP-xylose synthetase, *Journal of Biological Chemistry* **284**, 2576–2583.

5.3   Because of the low sequence similarity between glycosyltransferases, comparison of overall sequence identity is not a very effective way to search for new transferases in the genomes. Describe how alternative profile-based strategies, based on short, shared sequence motifs, can be used more effectively.

Reference: Narimatsu, H. (2006). Human glycogene cloning: focus on β3-glycosyltransferase and β4-glycosyltransferase families, *Current Opinion in Structural Biology* **16**, 567–575.

5.4   One of the difficulties in understanding how glycosyltransferases are held in the correct position in the Golgi apparatus is the absence of conventional retention signals that can be recognized by coat-specific adapter proteins. Discuss recent evidence that novel signals may be present in these enzymes.

Reference: Tu, L.T., Tai, W.C.S., Chen, L., and Bandfield, D.K. (2008). Signal-mediated dynamic retention of glycosyltransferases in the Golgi, *Science* **321**, 404–407.

5.5   Discuss possible reasons why amino acid swapping can convert a blood group A transferase into a blood group B transferase, but the reverse swapping experiment does not work.

5.6   By exploring the specificity of the relevant plant lectins and the defects in the relevant cell lines, explain (a) why Lec1 cells are resistant to ricin, phytohaemagglutinin and *Lens culinaris* lectin, but are more sensitive to treatment with concanavalin A than wild type cells are and (b) why Lec2 cells remain sensitive to ricin but are resistant to wheat germ agglutinin.

6

Glycomics and analysis of glycan structures

LEARNING OBJECTIVES

By the end of this chapter, students should understand:

1   How NMR, mass spectrometry, chromatography, and glycosidases are used to characterize glycan structures

2   The concept of glycomics and how data on glycan compositions of cells and tissues are being accumulated and made available in databases

3   Strategies for making synthetic glycans and their potential uses

The total complement of glycans in a cell or organism is referred to as the glycome. Because the expression of individual glycosyltransferases varies between cell types and at different stages in development, the glycomes of cells differ. Therefore, like the proteome, the glycome is a characteristic of a particular cell type at a specific state of differentiation. While genomes can be said to define organisms, proteomes and glycomes define cells. In addition, glycosylation varies between individuals, as demonstrated by the ABO blood group polymorphisms, and some variation is associated with disease conditions. Thus, analysis of glycomes provides a basis for understanding the functions of glycans in cell differentiation and disease. On the other hand, analysis of glycans on specific glycoproteins and glycolipids is needed to establish their roles as targets for receptor binding. Analysing the covalent structures of glycans attached to glycoproteins and glycolipids (glycan sequencing) is a more daunting task than the sequencing of proteins and nucleic acids. One difficulty lies in the branched nature of the structures and another is the extensive heterogeneity of the glycans, which means that a profile of glycans on a glycoprotein must be established rather than a single structure. This chapter provides a brief survey of the most commonly used methods for glycan structure determination and oligosaccharide synthesis to give some flavour of the power and the limitations of the strategies that are available.

6.1 NMR provides definitive information on oligosaccharide structures

Nuclear magnetic resonance (NMR) has been used to determine the structures of a large number of oligosaccharides. NMR analysis of oligosaccharides requires substantially more material than the mass spectrometric and chromatographic methods discussed in the next sections and it is less well-adapted to work on mixtures. However, it is able to provide information that cannot be obtained from other approaches and NMR serves as the ultimate source of definitive information on glycan structures.

This knowledge has been essential in defining the patterns of fragmentation observed in mass spectrometry and the activities of the glycosidases which underlie the other methods. Although NMR is well suited to the study of oligosaccharides, because they are small and are soluble at high concentrations, interpretation of the NMR spectra of sugars is made difficult by the similar environments of many of the protons. In the pyranose form of galactose, for instance, each of C2, C3, C4, C5, and C6 is bonded to a single oxygen atom, so the protons on these carbons are in very similar environments and hence have similar chemical shifts. However, the disposition of protons on adjacent carbons strongly affects the coupling between them, and because the different sugars have unique patterns of axial and equatorial protons around the ring, they can be distinguished (Figure 6.1). In the pyranoses, C1 is bonded to two oxygens, so the anomeric proton attached to this carbon is in a distinctive environment and is well resolved from the other protons. The chemical shift of this proton and its coupling to the proton on C2 are both different in the α and β anomers, so the NMR spectrum provides critical information on the stereochemistry of the linkages between sugars. With suitable adaptations, NMR can also be used to provide information about repeated linkages in polysaccharides as well as in lower molecular weight oligosaccharides.

6.2 Glycosidases can be used to analyse structures of glycans

Multiple methods are now available for determining the covalent structures of at least the more common types of protein- and lipid-linked glycans. N-linked glycans are released from polypeptides by treatment with anhydrous hydrazine or by digestion with peptide N-glycanase (PNGase) (Figure 6.2). Similar methods are available for the release of O-linked glycans by treatment with strong base and headgroups of glucosylceramides using glucosylceramidase. Following release, the pool of glycans can be separated into individual components so that individual glycans can be analysed by sequential degradation from the non-reducing termini using glycosidases of known specificity. The effect of the enzymatic treatments can be monitored by chromatography of the digestion products. In this approach, the progress of the digestions is followed by labelling the reducing end of the oligosaccharide with a fluorescent dye (Figure 6.2). The presence of the label provides a fixed point of reference, because released monosaccharides can be ignored and only the fragment containing the labelled reducing end

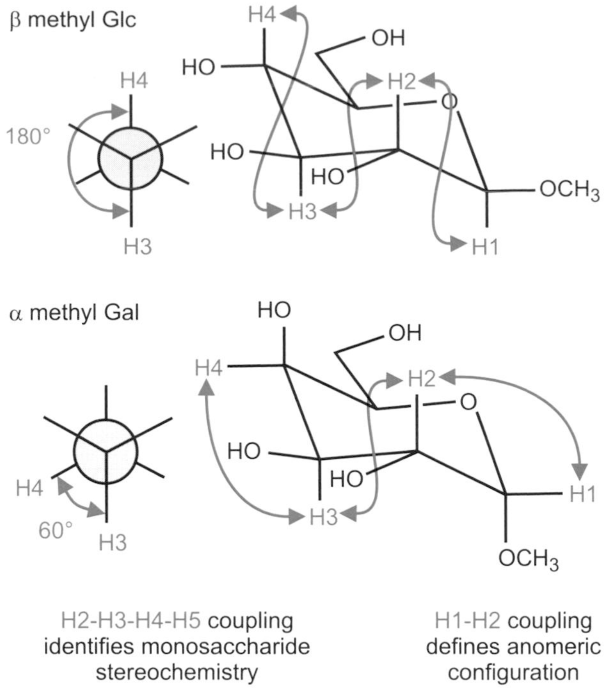

Figure 6.1 Use of NMR to determine monosaccharide and linkage configurations. Interactions of hydrogen nuclei on adjacent carbons depend on their relative orientation. Stronger coupling is observed when the hydrogens are in a *trans* arrangement (torsion angle 180°), while much weaker coupling is observed when they are in a *cis* arrangement (torsion angle 60°). Thus, the strength of coupling constants can be used to determine the disposition of hydrogens, and hence of hydroxyl groups characteristic of each monosaccharide. Similarly, the coupling of the H1 and H2 protons can be used to deduce the stereochemistry of a glycosidic linkage.

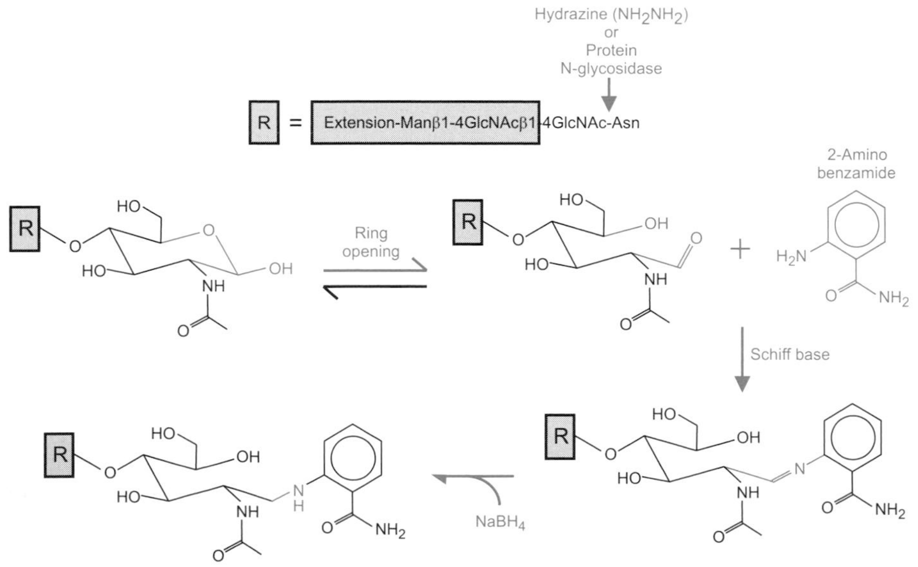

H2-H3-H4-H5 coupling identifies monosaccharide stereochemistry

H1-H2 coupling defines anomeric configuration

Figure 6.2 Release of N-linked glycans from proteins and labelling of the reducing end with the fluorescent tag 2-aminobenzamide. The released, labelled pool of glycans is separated into individual components by chromatography and the structure of each is determined.

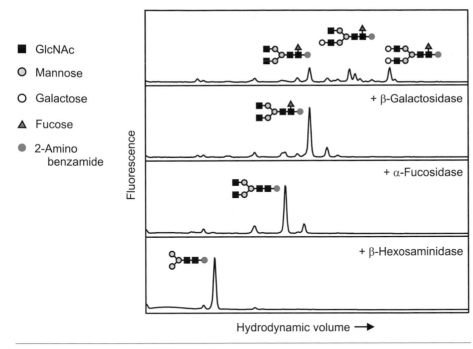

Figure 6.3 Sequential enzyme digestion of an end-labelled glycan. Gel filtration reveals that the initial pool of glycans is heterogeneous due to variable levels of terminal sialylation and galactosylation. After removal of these terminal structures, the core structure behaves as a single species. The glycans shown are from IgG, and are typical examples of bi-antennary complex N-linked oligosaccharides seen on serum glycoproteins. [Data from Alison Critchley, Tony Merry, and Pauline Rudd, University of Oxford.]

will be detected. Changes in elution position following digestions can be correlated with removal of a specific number of monosaccharides based on their known contribution to the chromatographic behaviour of the oligosaccharides (Figure 6.3). Lack of digestion with certain enzymes can provide important information.

For example, the neuraminidase (sialidase) from *Arthobacter ureafaciens* releases sialic acid in either α2-3 or α2-6 linkage, while the neuraminidase from Newcastle disease virus does not act on α2-6-linked sialic acid. Failure of the neuraminidase from Newcastle disease virus to digest an oligosaccharide that can be digested with the neuraminidase from *Arthobacter ureafaciens* suggests that the galactose residues are capped with α2-6-linked sialic acid residues.

There are many variations on the basic methodology for sequential degradation. Given that certain of the core elements are common to N-linked oligosaccharides, it is not always necessary to complete the digestion down to the reducing end sugar, because the chromatographic behaviour of the core structure has been well characterized and it can be recognized from its elution position. Indeed, in many cases the glycan composition of a pool of oligosaccharides can be deduced with reasonable accuracy from the chromatographic profile of the initial mixture. Sequential digestion can also be combined with internal labelling, in which radioactive sugars are fed to cells. For example, labelling with ³H-fucose can provide evidence for the presence of fucose, and release of the radioactivity following digestion with a fucosidase specific for the α1-6 linkage to *N*-acetylglucosamine then provides information on where the fucose is found in the glycan.

Analogous approaches can also be used to study other types of glycosylation, including the chemically quite different glycosaminoglycan chains of proteoglycans. In the case of heparan sulphate, labelling of the reducing end sugar is followed by an initial limited chemical cleavage with nitrous acid, which partially cleaves the chain at each sulphated glucosamine residue but does not cleave at unmodified GlcNAc residues. The lengths of the resulting fragments measured on polyacrylamide gels define the sites of *N*-sulphation. Shifts in the bands following treatment with specific sulphatases and exoglycosidases can then be used to determine the sequences working from the newly exposed non-reducing ends.

6.3   Mass spectrometry is particularly useful for analysis of complex mixtures containing small amounts of glycans

Mass spectrometry can be used both as a means to resolve the components of a glycan mixture and to derive information about the structures of individual glycans. Profiling of glycan mixtures is a particularly important approach to the analysis of glycomes of cells and tissues.

There are several different ways in which mass spectrometry can be used to analyse individual glycan structures. In the first instance, accurate masses determined by mass spectrometry of intact glycans are used to derive their compositions, because only a limited number of combinations of monosaccharide building blocks are consistent with the observed masses. Alternative mass spectrometric methods are used to produce fragments of the glycans directly in the mass spectrometer. The differences between the masses of the fragments provide information about the sequence of the glycan (Figure 6.4). Characteristic fragmentation patterns can also be used to distinguish sugars that have the same mass, such as mannose and galactose, and also make it possible to deduce which hydroxyl groups are involved in linkages. In combination with mass spectrometry, permethylation analysis can also be used to determine which hydroxyl groups on an oligosaccharide are involved in glycosidic linkages (Figure 6.5).

Structural analysis can be performed either on purified glycans or individual components of a mixture. In the latter case, tandem mass spectrometry allows initial separation of the mixture followed directly by fragmentation and re-analysis to obtain structural information. Alternatively, mass spectrometry can be used in place of chromatography as a sensitive way to analyse unlabelled glycans following sequential degradation with enzymes. These results can be interpreted in the same way as the chromatograms described in the preceding section. Finally, mass spectrometry is particularly useful for identification of unusual and unexpected modifications of glycans, including non-sugar substituents such as sulphate groups.

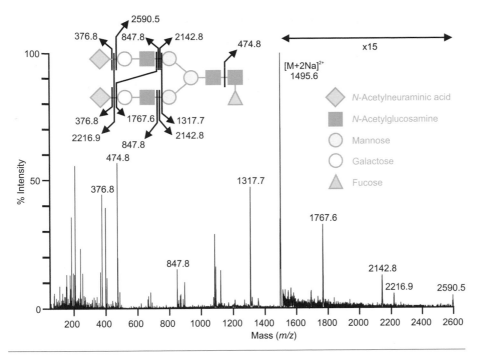

Figure 6.4 Fragmentation of a glycan observed in mass spectrometry. A single glycan, in this instance of mass 2590.5, can be isolated by mass spectrometry and immediately subjected to fragmentation and re-analysis by mass spectrometry (MS/MS analysis). Analysis of the fragmentation pattern allows deduction of many aspects of the structure. [Data from Simon North and Anne Dell, Imperial College, London.]

When there are multiple glycosylation sites in a single protein, information about which glycans were attached at which sites will be lost when glycans are released from the polypeptide. The glycans attached to different asparagine residues may be quite different. It is not uncommon for one asparagine to be conjugated only to high mannose oligosaccharides while others in the same protein would be conjugated to hybrid or complex structures. This information is usually established by **site analysis,** in which proteolytic cleavage of the polypeptide is followed by separation of the peptides and glycopeptides. Glycans released from the glycopeptides are then analysed separately to determine the distribution of glycans associated with each peptide. Alternatively, mass spectrometry can be used to analyse the peptide–glycopeptide mixture directly.

6.4 Glycomics provides a description of the glycans present in cells and tissues and is used to study receptors that bind glycans

Strategies for analysis of glycomes are being developed to track changes associated with development and disease. Mass spectrometry is particularly useful for profiling

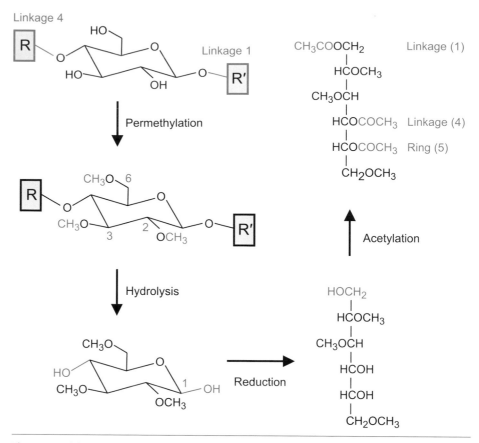

Figure 6.5 Linkage analysis using permethylation. All free hydroxyl groups in the intact glycan are methylated before the glycan is hydrolysed to its constituent monosaccharides and reduced to form a sugar alcohol. Hydroxyl groups that have been exposed by hydrolysis of the glycosidic linkages and opening of the ring structure are acetylated. The resulting mixture of volatile sugar derivatives is resolved by gas chromatography coupled to mass spectrometry. The fragmentation pattern for each constituent sugar can be used to identify the positions of the different modifications and hence of the linkages.

entire classes of glycans, such as all of the N-linked, O-linked, or glycolipid-linked glycans in tissues and in specific cell types (Figure 6.6).

Annotation of spectra is based on a combination of information about possible monosaccharide compositions to account for the observed masses and the known biosynthetic pathways for each class of glycan. Such annotation can now be achieved in a largely automated fashion. With improvements in sample preparation, it is possible to envision high-throughput screening of hundreds and even thousands of samples in the near future. Although a profile of this type does not fully define each of the proposed structures, it provides a global view of glycosylation. Ambiguity in assignment of certain structures can be resolved in follow-up experiments using tandem mass spectrometry on selected peaks. An initial goal of these

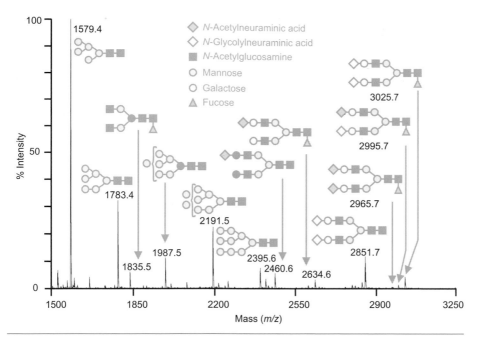

Figure 6.6 Glycan profiling by mass spectrometry. The major N-glycans released from mouse liver are resolved by mass spectrometry. The structures assigned to each peak are based on the monosaccharide composition consistent with the observed mass and by fragmentation data (Figure 6.4). [Data from Simon North and Anne Dell, Imperial College, London.]

screening approaches is to determine the degree of variation in glycomes between individuals and to track changes of glycosylation during cellular differentiation. Attempts are also being made to correlate differences in glycosylation with disease, for instance in various types of cancer.

⊖ See sections 13.7 and 13.8 for more on glycosylation in cancer.

6.5 Glycomic and genomic analysis provide some indication of the total size of the glycome in mammals

Sequence comparisons of glycosyltransferases and other enzymes required for glycan biosynthesis have defined motifs that have been used to identify the genomic complement of these glycogenes. Estimates of the total number of genes encoding proteins involved in glycan biosynthesis would be approximately 200 glycosyltransferase genes, 30 nucleotide sugar transporter genes, 20 processing glycosidase genes, and 50 genes for enzymes involved in precursor biosynthesis, with additional genes required for modifications such as sulphation. The overall complexity of the biosynthetic machinery is remarkable, given that the total of more than 300 genes would represent in excess of 1% of the genes in the human genome.

⊖ See sections 5.1 and 5.4 for more on sequence motifs in glycosyltransferases.

At the moment, glycosyltransferases have been identified that can explain most known glycan structures. However, the roles of a substantial number of the predicted glycosyltransferases remain to be defined. Some of these enzymes may be involved in the biosynthesis of structures present at abundances too low to be detected by current glycomic strategies. The extent of diversity in N-linked oligosaccharides can be estimated from structural analysis and from our knowledge of their biosynthesis. The fact that more than 500 different N-linked structures have already been characterized in chemical detail suggests that there may be as many as a thousand different types of glycans attached to asparagine residues in mammalian glycoproteins.

6.6   Systems glycobiology aims to link glycogene expression with glycosylation phenotypes

Correlations of gene expression data with glycomic analysis suggest that the major point of control of glycosylation is transcription of the relevant glycosyltransferases. A long-term aim is to monitor expression of the enzymes that synthesize glycans in parallel with analysis of the resulting glycomes of cells and tissues, with the goal of being able to predict the glycome from the pattern of expression of the biosynthetic enzymes. Achieving such predictive capability remains a substantial challenge, but it has the potential to define the flow of information from genes to glycosyltransferases to glycomes. Systems biology approaches to analysis of the data are being developed with some success, although the lack of large datasets of glycomic and gene expression data on well-defined cells limits the ability to make and test models. The possibility of control at points other than transcription will further complicate these efforts.

6.7   Glycan arrays can be used to define target ligands for glycan-binding proteins

see sections 6.10 and 6.11 for more on the synthesis of glycans

See sections 9.7 and 9.8 for more on DC-SIGN.

Information provided by glycomic strategies also provides a basis for detecting novel lectin-type recognition systems for glycans and for defining the targets of known glycan-binding receptors. Knowledge of the complement of potential protein- and lipid-linked glycans on cell surfaces and circulating glycoproteins forms the basis for construction of glycan arrays consisting of glycans purified from natural sources or synthesized by chemical and enzymatic methods immobilized on solid phase supports such as glass or polystyrene. The arrays are probed with glycan-binding receptors to identify the targets of the receptors (Figure 6.7). Comparison of binding to related oligosaccharides on such arrays provides an efficient means of determining which parts of glycans form the epitope recognized by the receptors. Screening of a glycan-binding protein on an array provides a relatively complete picture of the binding specificity of the

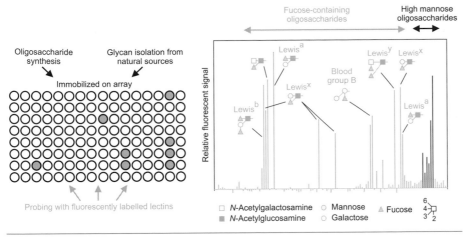

Figure 6.7 Binding selectivity of a glycan-binding receptor determined using a glycan array. Fluorescently labelled human DC-SIGN (see Chapter 9) binds to glycans immobilized in individual wells on an assay plate. Binding to two classes of glycans is evident: in addition to high-mannose oligosaccharides, glycans terminating in selected structures with terminal fucose and galactose or GalNAc is evident. [Adapted from: Guo, Y., Feinberg, H., Conroy, E., Mitchell, D.A., Alvarez, R., Blixt, O., Taylor, M.E., Weis, W.I., and Drickamer, K. (2004). Structural basis for distinct ligand-binding and targeting properties of the receptors DC-SIGN and DC-SIGNR, *Nature Structural and Molecular Biology* **11,** 591–598.]

protein, which can then be correlated with structural information to understand why some ligands bind with high affinity and others are excluded.

Because studies employing glycan arrays are leading to a better understanding of the biological roles of glycans, they are sometimes referred to as **functional glycomics.** Identification of ligands for receptors, coupled with knowledge of where and when such glycans are expressed, can provide important clues about the functions of the receptors. Arrays prepared with glycans from mammalian glycoproteins and glycolipids can be probed with mammalian receptors to study their roles in cell–cell interactions and glycoprotein trafficking, but such arrays can also be probed with receptors from bacteria and viruses to learn how these pathogens bind to and infect their hosts. Conversely, arrays that present oligosaccharides from the surfaces of bacteria can be probed with sugar-specific receptors of the mammalian innate immune system to learn how they recognize pathogen surfaces.

● See **section 13.8** for use of glycan arrays to detect cancer biomarkers.

6.8   Databases for glycobiology are being developed

Information about glycan structures is available in several different online databases, but no single repository has been established analogous to the international DNA and protein databases. Table 6.1 provides a selected set of databases that summarize information on glycomics, enzymes that synthesize and degrade oligosaccharides,

Table 6.1 Databases for glycobiology

Type	Link
Glycan structures and analysis	
CFG: Glycomics database	www.functionalglycomics.org >CFG databases > Glycan Structures
SweetDB: NMR and mass spectrometry data	www.glycosciences.de > Databases
Glycan biosynthesis	
KEGG: glycan biosynthetic pathways	www.genome.jp >KEGG PATHWAY > Metabolism [1.7 Glycan Biosynthesis and Metabolism]
CAZy: carbohydrate-active enzymes	www.cazy.org
Glycan-binding receptors	
Animal lectin resource	www.imperial.ac.uk/research/animallectins/
CFG: Glycan-binding protein database	www.functionalglycomics.org/CFGparadigms

Notes: Links accessed November 2010.

and glycan-binding proteins. Most of the existing databases represent retrospective efforts to capture published information on glycans attached to individual glycoproteins, a process that requires a substantial degree of curation to ensure accuracy and completeness.

With the increasing volume of glycomic data being generated, the databases are evolving to accept information on cells and tissues. Mechanisms are also being developed to ensure that new data are deposited directly, as they are for DNA sequences and protein structures. The databases serve different purposes for different users, ranging from the ability to locate information on where particular types of glycans have been found to more specific experimental results used to establish the structures. Rather than developing a single, centralized database, it is envisioned that existing databases will be enlarged and linked to each other, to take advantage of their individual strengths and different data manipulation capabilities.

In addition to information on glycans themselves, further databases for the enzymes that synthesize and degrade them, and for the glycan-binding proteins, are being developed. Coverage of the human glycosyltransferases in the databases is relatively complete and allows identification of structural and evolutionary relationships between families of glycosyltransferases and glycosidases. A major goal is now to establish relational databases that link the properties of these enzymes to the information on glycan structures. For example, it will eventually be possible to use such relational databases to see which enzymes are required for synthesis of any particular glycan in the human glycome. Similarly, databases of animal lectins provide a nearly complete description of the families of receptors and some of their ligand-binding properties, but data from the screening of glycan arrays will make it possible to link the receptor databases with the appropriate targets in the glycomics databases.

6.9   Glycoconjugates in cells and tissues can be analysed using lectins

Plant or animal lectins that bind specifically to certain oligosaccharides can be used to identify specific sugar structures. Plant lectins have been particularly useful in this respect, since they often bind with high selectivity to a specific oligosaccharide sequence within a glycan. For example, binding of a plant lectin that recognizes the A blood group trisaccharide to a glycoprotein or glycolipid can be taken as evidence for the presence of the blood group sequences in the glycan portions of the glycoconjugate.

The ability of lectins to detect specific glycan structures in intact glycoconjugates makes them useful in staining mixtures of glycoproteins separated by polyacrylamide gel electrophoresis, or mixtures of glycolipids resolved by thin layer chromatography. In addition, lectins can be used to localize specific glycans in cells by conjugating the lectins to probes, such as fluorescent tags, that can be visualized by microscopy (Figure 6.8). Antibodies that bind specific glycan sequences can be used in similar ways. In this way, lectins and anti-carbohydrate antibodies form

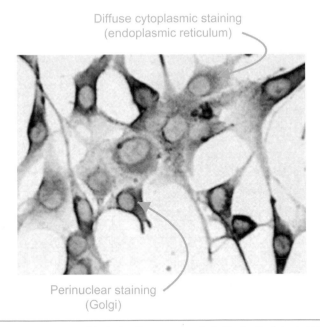

Diffuse cytoplasmic staining
(endoplasmic reticulum)

Perinuclear staining
(Golgi)

Figure 6.8 Staining of fibroblasts with the lectin concanavalin A. The lectin detects N-linked glycans, which are present on glycoproteins in the endoplasmic reticulum and Golgi compartments. [Modified from: Tian, E, Ten Hagen, K.G., Shum, L., Hang, H.C., Imbert, Y., Young, W.W., Jr., Bertozzi, C.R., and Tabak, L.A. (2004). An inhibitor of O-glycosylation induces apoptosis in NHI3T3 cells and developing mouse embryonic mandibular tissues, *Journal of Biological Chemistry* **279**, 50382–50390, with permission.]

important bridges between the chemical analysis of glycan structures and the biological study of their functions.

6.10   Small oligosaccharides can be synthesized using chemical methods

The limited quantities of biological macromolecules obtained from natural sources often make these molecules inaccessible to studies of conformation and function. Chemical synthesis of oligosaccharides can provide the quantities of homogeneous glycans needed for such studies.

There are two major challenges in oligosaccharide synthesis. First, only a single anomer must be produced during the coupling step. In the absence of chemical methods that yield complete anomeric specificity, the mixture of α and β anomers must be separated and purified, which vastly increases the effort and decreases yields.

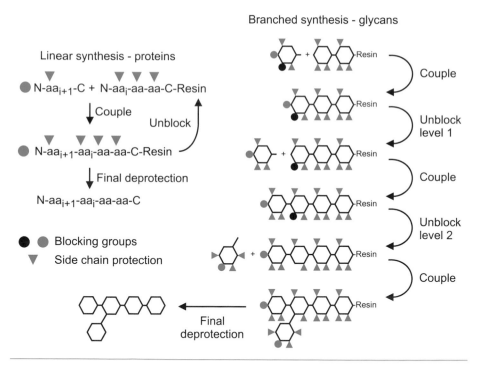

Figure 6.9 Comparison of strategies for chemical synthesis of peptides and oligosaccharides. In peptide synthesis, the carboxyl group of the C-terminal amino acid is conjugated via a stable linkage to a resin to facilitate washing away of reagents between reaction steps. Stable side chain protecting groups remain on throughout the synthesis, while the amino group of each incoming amino acid bears a labile blocking group that can be removed to allow coupling of the next residue. Because many glycans are branched, there must be two or more levels of blocking group stability, so that one hydroxyl group can be made free to react, creating one branch, and then the second can be released and reacted to form a second branch.

This requirement also means that synthesis on a solid phase support cannot be used. Second, the use of a hierarchy of chemical stability developed for synthesis of linear peptides and nucleic acids must be adapted for synthesis of branched oligosaccharides (Figure 6.9). Custom-made precursors with blocking and protecting groups on different hydroxyl groups are needed to create different branched structures. In spite of these difficulties, synthetic schemes have been established for some biologically important but relatively small oligosaccharides.

⊙ See section 13.8 for information as oligosaccharides as cancer vaccines.

6.11   Enzymes provide an alternative method for the synthesis of oligosaccharides

Enzymes provide a useful alternative to chemical synthesis of oligosaccharides. Cloned glycosyltransferases of known specificity can be expressed and used to create desired linkages. A potential limitation of this method is the need for nucleotide sugars to be used as donors at each step. However, additional enzymes can be included in the synthetic scheme to regenerate nucleotide sugars from relatively inexpensive starting materials (Figure 6.10). Enzyme cycles have been established to generate many of the sugar nucleotide donors needed for the synthesis of a variety of oligosac-

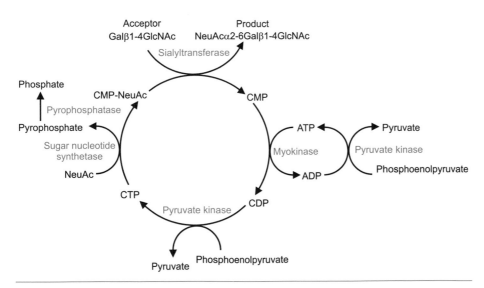

Figure 6.10 An example of a regeneration cycle that can be used to add sialic acid to a glycan terminating in an *N*-acetyllactosamine unit. This type of sialylated structure is found in complex N-linked oligosaccharides. Similar cycles can be used for other steps in glycan synthesis, and in some cases, unnatural sugar analogues can be incorporated using the same strategy. ADP, adenosine diphosphate; ATP, adenosine triphosphate; CDP, cytidine diphosphate; CMP, cytidine monophosphate; CTP, cytidine triphosphate. [Adapted from: Sears, P. and Wong, C.-H. (1996). Intervention of carbohydrate recognition by proteins and nucleic acids, *Proceedings of the National Academy of Sciences U.S.A.* **93**, 12086–12093.]

See section 13.5 for more on uses of glycosyltransferases in biotechnology.

charides. With the availability of an increasing number of cloned glycosyltransferases, many additional biologically interesting sugars should become accessible to synthesis in the future.

Peptide and oligonucleotide synthesis have been used to create analogues of natural molecules that differ in subtle ways, such as substitutions of single amino acid residues or individual bases. The ability to create similar changes in glycans would

---

BOX 6.1 Box 6.1 Glycotherapeutics: *Synthetic heparin oligosaccharides are used to control blood clotting*

Inadvertent or prolonged activation of the protease activation cascade leading to blood coagulation can result in inappropriate formation of blood clots. Clots released into the blood stream can lodge in veins of the legs or lungs, causing deep vein thrombosis or pulmonary embolisms. Antithrombin III is a key regulator of the final steps in the pathway leading to generation of the clot-forming fibrin, because it inhibits the activated forms of thrombin (Factor IIa) and the protease that activates thrombin, Factor Xa. Heparin sulphate attached to proteoglycans at the surface of vascular endothelial cells (Chapter 3) facilitates the interaction of antithrombin III with thrombin and Factor Xa, preventing clot formation in unperturbed parts of the circulatory system. Heparin, a heavily sulphated form of heparan isolated from the mucus-secreting cells of pig intestine, has long been used to prevent clot formation. Treatment with heparin is particularly effective in preventing thrombosis in patients undergoing hip and knee replacement surgery.

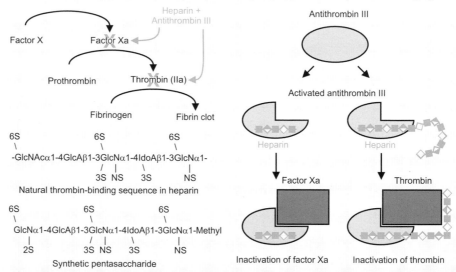

In overall composition, heparin resembles the heparan sulphate attached to proteoglycans on cell surfaces, but it is particularly effective as an activator of antithrombin III. Fragmentation of heparin into oligosaccharides, followed by isolation of fragments that can activate antithrombin III, led to identification of a specific pentasaccharide sequence in heparin that binds to antithrombin III. Binding causes a conformational change in antithrombin III, which in turn leads to

Box 6.1   101

exposure of an arginine residue that binds to the active sites of the Factor Xa and thrombin pro-teases. The importance of individual modifications to the heparan sulphate backbone within the pentasaccharide has been demonstrated by preparing synthetic oligosaccharides containing the natural structure and modified versions. Although the pentasaccharide is sufficient to facili-tate interaction of antithrombin III with Factor Xa, inhibition of thrombin requires longer heparin oligosaccharides. In addition to the pentasaccharide that binds to antithrombin III, these oligo-saccharides contain a further negatively charged sequence that interacts in a relatively non-specific way with a positively charged patch on the surface of thrombin, thus stabilizing the thrombin–antithrombin III complex.

Because heparin prepared from natural sources is subject to batch-to-batch variability and potential contamination with infectious agents, chemically synthesized antithrombin III-binding pentasaccharide has been developed as an alternative therapeutic. The synthetic material cur-rently licensed for clinical use is nearly identical to the natural pentasaccharide. This drug lacks the secondary thrombin-binding sequence and therefore only activates antithrombin III binding to Factor Xa. However, inhibition of Xa substantially reduces thrombin activation.

Synthesis of this compound from abundant starting materials such as glucose, glucosamine, and sulphuric acid requires more than 50 steps, so structurally similar but synthetically more acces-sible compounds that still activate antithrombin III are being developed as potential second gen-eration drugs. In addition, more complex compounds with secondary thrombin-binding sites have been explored as potential activators of antithrombin III inhibition of thrombin. Such compounds might have therapeutic potential in a broader range of clinical situations where inhibition of thrombin as well as Factor Xa is required. However, alternative chemical–enzymatic approaches may be required to make sufficiently large molecules.

Essay topics

- Describe the analytical and synthetic approaches that have been used to define the antithrombin III-binding pentasaccharide fragment of heparin.

- Discuss how the dual roles for heparin in antithrombin III activation and bridging to thrombin have been demonstrated and how they might be exploited in further stages of drug development.

Lead references

- Bauer, K.A. (2003). New pentasaccharides for prophylaxis of deep vein thrombosis, *Chest* **124**, 364S–370S.

- Johnson, D.J.D., Langdown, J., and Huntington, J.A. (2010). Molecular basis of factor IXa recognition by heparin-activated antithrombin revealed by a 1.7-Å structure of the ternary complex, *Proceedings of the National Academy of Sciences U.S.A.* **107**, 645–650.

- Laremore, T.N., Zhang. F., Dordick, J.S., Liu, J., and Linhardt, R.J. (2009). Recent progress and applications in glycosaminoglycan and heparin research, *Current Opinion in Chemical Biology* **13**, 633–640.

- Petitou, M. and van Boeckel, C.A.A. (2004). A synthetic antithrombin binding pentasaccharide is now a drug: what comes next?, *Angewandte Chemie, International Edition* **43**, 3118–3133.

also be useful. It might seem that the chemical methods would be more adaptable to the creation of non-natural oligosaccharides, but at least some glycosyltransferases can be used to add novel sugars (given the right donors). The ability to create novel linkages may require modification of the transferases, which might become possible as our understanding of their structures becomes more extensive.

6.12   Neoglycoconjugates can be created by chemically linking sugars to proteins or lipids

Many of the biological effects of glycans are only evident in the context of the proteins to which they are attached. For these reasons, it would be useful to be able to conjugate modified glycans to proteins in a site-specific way to create glycoproteins with truly defined structures. Chemical methods for creating such neoglycoproteins have been developed. For example, simple monosaccharides and oligosaccharides can be attached to unglycosylated carrier proteins such as serum albumin. Neoglycoproteins created in this way have been very useful in the identification and characterization of lectins. One particular advantage of this approach is the possibility of controlling the density of sugar ligands.

The linkages of glycans to proteins in chemically generated neoglycoproteins are usually through lysine or cysteine side chains. No chemical method exists to attach glycans to specific asparagine residues in the way they are found in natural glycoproteins. Therefore, glycoproteins with defined N-linked glycans must be prepared using biosynthetic pathways. Addition of inhibitors of specific steps in the glycosylation pathway can be used to generate unique glycan structures. A similar effect can be achieved by the production of proteins in mutant cell lines that lack specific glycosyltransferases. It is also possible to modify existing structures *in vitro* using glycosidases to remove some of the sugars and glycosyltransferases to rebuild alternative glycans.

⊛ See section 5.7 for more on mutant cell lines.

⊛ See section 13.5 for more on use of enzymes for making alternative glycans.

SUMMARY

Because common sets of glycans are attached to many glycoproteins, the structures that are present can often be identified through routine profiling, with confirmation of particular structures being achieved by enzyme digestion and mass spectrometric fragmentation. Lectins and antibodies can also be used to detect specific sugar structures and are particularly useful for localizing glycans in cells and tissues. Recent advances in glycan profiling are also providing insights into the glycomes of cells and tissues. Experimental and bioinformatics approaches are being developed to link this information to knowledge about the enzymes that synthesize the glycans and the glycan-binding receptors that recognize them. Synthesis of glycans, although still an evolving field, will probably play an increasing part in analysing conformations and functions of glycans.

KEY REFERENCES

Bartolozzi, A. and Seeberger, P.H. (2001). New approaches to the chemical synthesis of bio-active oligosaccharides, *Current Opinion in Structural Biology* **11**, 587–592. Advances in chemical synthesis of oligosaccharides are reviewed here and by Wong (2005) (see below).

Dell, A. and Morris, H.R. (2001). Glycoprotein structure determination by mass spectrometry, *Science* **291**, 2351–2356. Methods of structure determination by mass spectrometry are illustrated with applications in glycobiology.

Drickamer, K. and Taylor, M.E. (2002). Glycan arrays for functional genomics, *Genome Biology* 3, 1034.1–1034.4. This commentary provides an overview of strategies for developing gly-can arrays.

Duus, J.Ø., Gotfredsen, C.H., and Bock, K. (2000). Carbohydrate structural determination by NMR spectroscopy: modern methods and limitations, *Chemical Reviews* **100**, 4589–4614. This review provides a brief summary of traditional methods for analysing glycan struc-tures by NMR, followed by a selection of recent advances.

Dwek, R.A., Edge, C.J., Harvey, D.J., Wormald, M.R., and Parehk, R.B. (1993). Analysis of glyco-protein-associated oligosaccharides, *Annual Review of Biochemistry* **62**, 65–100. This arti-cle, and the one by Fukuda and Kobata (1993) below, provide details of the basic methods for analysing glycan structure.

Feizi, T. (1985). Demonstration by monoclonal antibodies that carbohydrate structures of glycoproteins and glycolipids are oncodevelopmental antigens, *Nature* **314**, 53–57. A clas-sic summary of how antibodies can be used to probe glycoconjugates.

Fukuda, M. and Kobata, A. (1993). *Glycobiology: A Practical Approach.* Oxford: Oxford University Press. See Dwek *et al.* (1993) above.

Nairn, A.V., York, W.S., Harris, K., Hall, E.M., Price, J.M., and Moremen, K.W. (2008). Regulation of glycan structures in animal tissues: transcription profiling of glycan-related genes, *Journal of Biological Chemistry* **283**, 17298–17313. An example of how glycogene expres-sion profiling is being undertaken.

Rudd, P.M. and Dwek, R.A. (1997). Rapid, sensitive sequencing of oligosaccharides from glycoproteins, *Current Opinion in Biotechnology* **8**, 488–497. Methods for the release, labelling, and sequencing of glycans are reviewed in detail.

Taylor, M.E. and Drickamer, K. (2009). Structural insights into what glycan arrays tell us about how glycan-binding proteins interact with their ligands, *Glycobiology* **19**, 1155–1162. A discussion of how results from screening of glycan arrays can be correlated with structur-al information on glycan-binding receptors.

Wong, S.Y.C. (1995). Neoglycoconjugates and their applications in glycobiology, *Current Opinion in Structural Biology* **5**, 599–604. An overview of the methods for making glyco-conjugates, and their experimental uses.

Wong C.H. (2005). Protein glycosylation: new challenges and opportunities, *Journal of Organic Chemistry* **70**, 4219–4225. See Bartolozzi and Seeberger (2001) above.

QUESTIONS

6.1   Which enzymes would be required for analysis of the structure of a sialylated complex tri-antennary oligosaccharide by sequential enzyme degradation and chromatography?

6.2   Explain why knowledge of the biosynthetic pathways for N-linked and O-linked glycans is essential when assigning structures to oligosaccharides analysed by mass spectrometry.

6.3   You are working on a cell surface receptor that has three potential N-linked glycosylation sites and a proline-rich sequence containing several threonine residues in its extracellular region. You have expressed a soluble form of the receptor in Chinese hamster ovary cells and can purify approximately 1 mg from the medium. What approaches would you use to characterize the glycosylation of the expressed form of the receptor?

6.4   Go to the web page for the Consortium for Functional Glycomics at **http://www.functionalglycomics.org/glycomics/common/jsp/firstpage.jsp** and open the Glycan Structure Database. Using search by multiple criteria, first search for glycans within the molecular weight range 1680 to 1720 (enter 1700 ± 20) and look at some of the structures. Then refine your search to find which glycans of this molecular weight are: a) N-linked; b) O-linked; c) N-linked complex; d) N-linked complex found in humans (enter homo for species). Why are there many more N-linked than O-linked glycans of this size?

6.5   Describe the methods that were used to analyse the structures of O-linked glycans on P-selectin glycoprotein ligand- 1.

Reference: Wilkins, P.P., McEver, R.P., and Cummings, R.D. (1996). Structures of the O-glycans on P-selectin glycoprotein ligand-1, *Journal of Biological Chemistry* **271**, 18732–18742.

6.6   Discuss how use of a glycan array in combination with other techniques can identify specific oligosaccharide ligands for sugar-binding proteins and provide insight into the biological functions of the proteins.

Reference: Blixt, O., Head, S., Mondala, T., Scanlan, C., Huflejt, M.E., Alvarez, R., Bryan, M.C., Fazio, F., Calarese, D., Stevens, J., Razi, N., Stevens, D.J., Skehel, J.J., van Die, I., Burton, D.R., Wilson, I.A., Cummings, R., Bovin, N., Wong, C.H., and Paulson, J.C. (2004). Printed covalent glycan array for ligand profiling of diverse glycan binding proteins, *Proceedings of the National Academy of Sciences U.S.A.* **101**, 17033–17038.

Bochner, B.S., Alvarez, R.A., Mehta, P., Bovin, N.V., Blixt, O., White, J.R., and Schnaar, R.L. (2005). Glycan array screening reveals a candidate ligand for Siglec-8, *Journal of Biological Chemistry* **280**, 4307–4312.

Coombs, P.J., Graham, S.A., Drickamer, K., and Taylor, M.E. (2005). Selective binding of the scavenger receptor C-type lectin to Lewis[x] trisaccharide and related glycan ligands, *Journal of Biological Chemistry* **280**, 22993–22999.

Conformations of oligosaccharides

LEARNING OBJECTIVES

By the end of this chapter, students should understand:

1   The definition of torsion angles used to describe glycan conformations
2   The nature of the interactions that determine glycan conformations
3   Experimental methods used to analyse glycan conformations
4   How linkages in polysaccharides confer on them useful large-scale conformations

A major goal in glycobiology is to relate structures of glycans to their functions. Glycans with different covalent structures take up different conformations in the same way that proteins with different sequences have distinct tertiary structures or folds. Sequences of proteins dictate their three-dimensional structures and these structures determine their functions. Similar rules probably apply in the case of oligosaccharides. However, principles relating structure to function have not been easy to establish, partly because of our limited knowledge of the conformations that glycans assume. This chapter summarizes what is known about the three-dimensional structure of N-linked oligosaccharides.

7.1   Three-dimensional structures of oligosaccharides are called conformations

Because oligosaccharides can be branched, it is often inappropriate to use the term sequence to refer to the way that hexose and hexosamine building blocks are connected. Therefore, the term structure is usually used to denote the covalent structure of glycans. As a result, the word structure cannot be used as shorthand for three-dimensional structure as it is in the case of DNA and proteins. To avoid confusion, the term conformation is more commonly used to refer to the arrangement of oligosaccharides in three-dimensional space.

7.2   Monosaccharides assume a limited number of conformations

Hexose and hexosamine residues in glycans are relatively rigid, but they can theoretically assume several different conformations (Figure 7.1). Chair conformations are generally favoured over boat conformations, because boat conformations cause crowding of the atoms in the ring. In the different chair conformations, each group attached to the ring either projects outward from the edge of the pyranose ring in an equatorial position, or points upward or downward from the ring in an axial

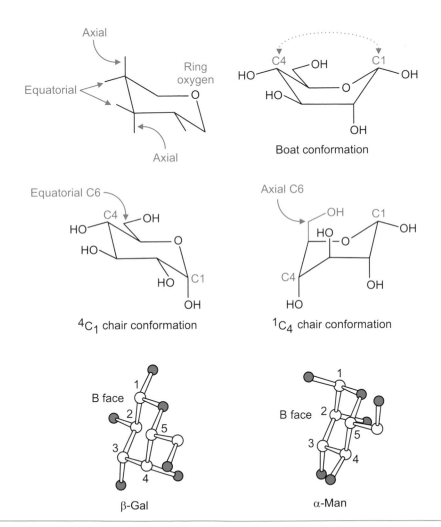

Figure 7.1 Monosaccharide conformations. Crowding between hydrogens on C1 and C4 makes the boat conformation less stable than the chair conformations. The $^4C_1$ chair conformation results in the least crowding, because the exocyclic C6 and the hydroxyl groups 2, 3, and 4 are in an equatorial position, while in the $^1C_4$ chair conformation they are axial. The apolar B face is defined as the face on which the carbons are numbered in an anticlockwise arrangement.

position. Atoms in the axial positions are more likely to clash sterically. In the hexose residues that predominate in biological glycans, one of the chair conformations, designated $^4C_1$, is energetically favoured because this arrangement places the bulky C6–O6 group and most of the hydroxyl groups in equatorial positions. In this conformation, one face of the ring is relatively non-polar. This is the B face for the commonly occurring hexoses.

● See **Section 10.7** for more on the importance of the non-polar B face.

The fact that pyranose rings behave as fixed units greatly simplifies description of the conformations of glycans. There are relatively few substituents and thus little of the side chain conformational variability that characterizes protein structures. However, rotation about the C5–C6 bond can occur (Figure 7.2). As with any single bond, **staggered** conformations are preferred because these maximize the distance between atoms attached to the two atoms at either end of the rotating bond. These rotations are described by the **torsion angle** ω. There is also rotational flexibility about the C2–N2 bond in the amino sugars.

7.3 Torsion angles are used to describe conformations of glycans

Because the conformations of the individual sugar residues are largely known, describing the conformations of oligosaccharides is mostly a matter of describing

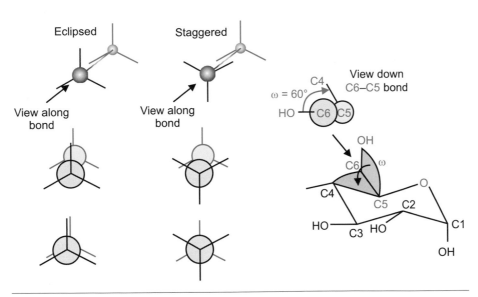

Figure 7.2 Representations of possible conformations generated by rotation about a single bond. A staggered conformation is preferred to maximize the distance between atoms. Three possible staggered conformations can be created by rotations of 120°. Newman projections provide a shorthand way to display the conformations. Rotational flexibility about the C5–C6 bond is described by the ω angle, which is measured between the plane defined by O6–C6–C5 (blue) and the C6–C5–C4 plane (grey). This angle can also be described as the O6–C6–C5–C4 torsion angle.

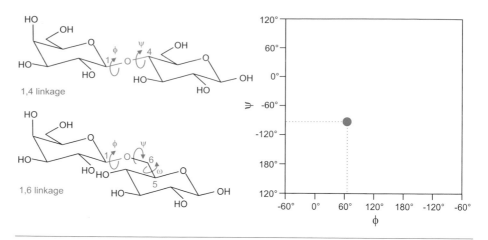

Figure 7.3 Torsion angles in glycosidic linkages and torsion angle plot. Two torsion angles, ϕ and ψ, are required to define the conformation of most linkages, while an additional ω angle is required to describe a 1-6 linkage. Thus, the conformations of most linkages can be described on a two-dimensional plot of ψ versus ϕ.

the linkages between residues. The torsion angles used to describe these linkages are analogous to the torsion angles used to describe the backbone conformation of polypeptides (Figure 7.3). There are two single bonds in the glycosidic linkage, so both the **torsion angle** ϕ and the torsion angle ψ are needed to describe the conformation. In the case of linkages to O6, the value of the ω angle will also affect the overall conformation of the linkage.

Because two angles suffice to describe the conformation of most glycosidic linkages, it is convenient to display the possible conformations on a two-dimensional plot of ϕ versus ψ, similar to the **Ramachandran plots** used to describe protein backbone conformations. Such plots allow ready comparison of the conformations of different linkages and are particularly useful when describing energetic constraints on conformations of glycans. Unfortunately, there is no absolute agreement on the conventions used to define the torsion angles, so care must be taken when comparing torsion angle plots from different sources. The conventions observed here are those that are used in crystallography.

7.4   Local steric and electronic interactions limit the possible conformations of glycosidic linkages

The first constraint that determines the conformation of a glycan is whether a particular combination of torsion angles leads to unacceptable steric clashes between two linked hexose or hexosamine residues. For peptide bonds, roughly three-quarters of the possible conformations are inaccessible because many combinations of torsion angles bring atoms on either side of the peptide bond too close together. These excluded conformations are detected by treating atoms as hard spheres defined by their van der Waals radii and noting which conformations can be achieved when the torsion angles are varied. For glycosidic linkages, these steric considerations do not rule out such a high proportion of the torsion angle combinations. There are, however, preferred values which are determined by local steric effects in combination with the electronic properties of the glycosidic linkage.

To illustrate the role of different local interactions, it is useful to start with the ϕ angle (Figure 7.4). Based on steric considerations, a torsion angle might be expected to have three possible preferred values, corresponding to the positions at which the substituents on the atoms at each end of the bond are staggered. For ϕ, these correspond to values of $-60°$, $+60°$, and $+180°$. Selection of the most stable among these possible values depends in part on the size of the substituents. The electronic effects on preferred torsion angles result from the fact that C1 in the pyranose form of the hexoses and hexosamines is flanked by two oxygen atoms, each with two sp^3-hybridized orbitals containing unshared pairs of electrons (Figure 7.5). The orientation of these two orbitals on the ring oxygen is fixed by the ring geometry, but the orientation of the orbitals on the glycosidic oxygen is described by the ϕ angle. The preferred ϕ value is the one that maximizes the distance between the unshared pairs of electrons, which corresponds to $+60°$ for a D-hexose in α linkage and $-60°$ for a D-hexose in β linkage. This phenomenon, known as the **exo-anomeric effect**, makes these values significantly more favourable energetically than other values and in practice these are by far the most

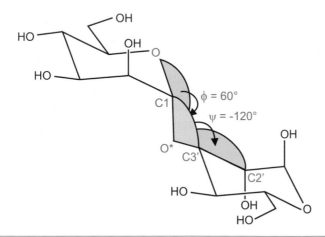

Figure 7.4 Definition of torsion angles. The oxygen atom that forms the glycosidic linkage is marked O*. The ϕ angle is measured between the O–C1–O* plane (blue) and the C1–O*–C4′ plane (grey). This angle is also referred to as the O–C1–O*–C4′ torsion angle. The ψ angle is measured between the C1–O*–C4′ plane (grey) and the O*–C4′–C3′ plane (blue) and, therefore, can be called the C1–O*–C4′–C3′ torsion angle.

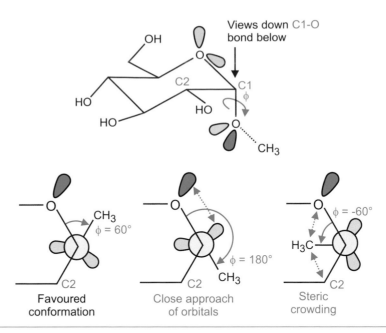

Figure 7.5 The exo-anomeric effect for an α linkage. A φ angle of −60° minimizes steric crowding and overlap of the unshared pairs of electrons in orbitals on the ring oxygen and the oxygen forming the glycosidic linkage.

commonly observed. Although the exo-anomeric effect is an experimentally demonstrated fact, the actual electronic basis underlying it is more complex than suggested by the orbital repulsion idea discussed here. The restriction on φ angles resulting from the exo-anomeric effect is like the restriction that a peptide bond in a polypeptide must be planar, because of electronic interactions involved in double bond formation.

Although there is no exo-anomeric effect for the ψ angle, steric considerations generally restrict the angle to only one of the three possible staggered conformations for a particular linkage. As noted above, for 1–6 linkages, the ω angle must also be considered. Of the three ω angles that give staggered conformations, one often produces a steric clash, because O5 and O6 are unacceptably close, so only two of the conformations will be observed, corresponding to ω = 60° and +180° (Figure 7.6).

Rather than considering the sugar atoms as hard spheres, it is possible to use more realistic representations of how they interact, with an energetically favourable interaction as two atoms approach each other followed by a negative interaction when they get too close. Similarly, the exo-anomeric effect does not restrict φ to a single value, but makes a different energetic contribution depending on how close to the optimal value φ is. Using these more sophisticated energetic considerations, the energy associated with different combinations of torsion angles can be calculated and presented in the form of a series of contours on a torsion angle plot

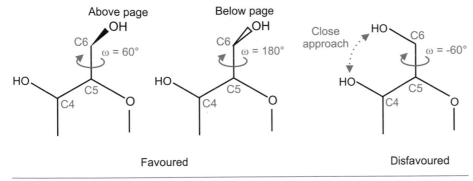

Figure 7.6 Conformational constraints on the ω angle. Of the three possible ω angles giving staggered conformations, one (ω = –60°) is disfavoured because it brings hydroxyl groups on C4 and C6 close together.

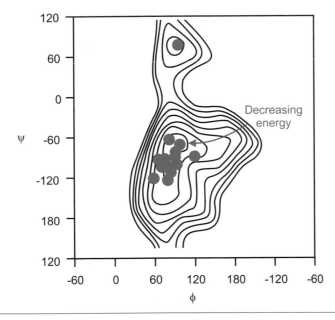

Figure 7.7 Energy contours for various Manα1-3Man disaccharides. Both primary and secondary energy minima are present. As indicated by the symbols, most of the experimentally observed conformations fall in the primary minimum, but the conformation associated with the secondary minimum has also been seen. [Adapted from: Imberty, A. (1997). Oligosaccharide structures: theory versus experiment, *Current Opinion in Structural Biology* **7**, 617–623.]

(Figure 7.7). Energy contour plots for glycosidic linkages reveal relatively shallow energy minima. There are often several local minima that differ little in energy. The observed torsion angles for various glycosidic linkages generally fall close to minima on the energy contours.

7.5   The conformation of an oligosaccharide is influenced by interactions between hexoses distant from each other in the covalent structure

Although the rigid hexose rings, the exo-anomeric effect, and steric factors mean that only a few conformations will be favoured for most linkages, the torsion angles are not fixed precisely by these considerations. Structural analysis shows that, for any particular linkage, torsion angles can often range over 30–40°. The cumulative effect of such small differences is to generate a diversity of overall conformations, even for identical glycans.

The overall conformations of oligosaccharides and polysaccharides are determined by local constraints at the glycosidic linkages, in combination with interactions between sugar residues that are distant from each other in the covalent structure. Some of these interactions can be unfavourable, when steric clashes between hexose residues that are separated by one or more intervening residues make some combinations of torsion angles impossible in an oligosaccharide. Electrostatic repulsion between negatively charged groups in glycans, such as carboxyl groups in sialic acid residues, can also influence conformation. On the other hand, favourable hydrogen bonds and hydrophobic interactions stabilize selected conformations of individual glycosidic linkages. The hydroxyl groups of the sugar residues are potential hydrogen bond donors and acceptors and the ring oxygen is a potential hydrogen bond acceptor. For the Manα1-2Man linkage, a hydrogen bond can form directly between the adjacent monosaccharide residues (Figure 7.8), while in other cases, hydrogen bonds involve intervening water molecules. Conformations can also be stabilized by van der Waals interactions, either between adjacent residues or those that are further apart (Figure 7.9). These associations reflect in part the amphipathic character of hexoses, in which the B face is often

See Sections 9.6 and 9.7 for more on functions of the Lewis* trisaccharide.

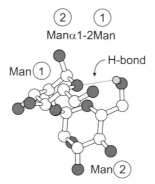

Figure 7.8 Stabilization of a specific conformation of a glycosidic linkage by hydrogen bonding. One conformation of the Manα1-2Man disaccharide is preferred because of the energetically favourable formation of a hydrogen bond between O6 of the non-reducing mannose residue and the ring oxygen of the reducing mannose residue. [Based on entry 3P5F in the Protein Databank.]

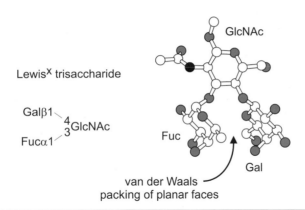

Lewis^X trisaccharide

Galβ1
          4
          3 GlcNAc
Fucα1

Figure 7.9 Hydrophobic packing between apolar faces of monosaccharides in a small oligosaccharide. The preferred conformation of the branched trisaccharide is stabilized by van der Waals contact between non-polar portions of the two non-reducing terminal residues. [Based on entry 2KMB in the Protein Databank.]

particularly hydrophobic because the polar hydroxyl groups are orientated either equatorially or facing upward on the opposite face. These interactions are analogous to those that form a hydrophobic core in a protein, but they can occur on a relatively local basis due in part to the rigidity and near planarity of the hexoses.

7.6 Co-operative interactions determine the overall folds of oligosaccharides

Although certain conformations are favoured for most glycosidic linkages, selection of particular angles from among the allowed choices is not fully determined by the sugars on the two sides of the linkage. An oligosaccharide assumes its final conformation as a co-operative event. A combination of torsion angles at multiple glycosidic linkages is selected because the sum of all the interactions produced by this combination is the most favourable. This principle is the same as that which determines the folding of proteins. However, the relatively small sizes of most oligosaccharides mean that the degree of co-operativity, reflecting the total number of favourable interactions that occur in the preferred conformation, is small compared with protein domains. Protein modules that have distinct folds generally consist of 50 or more amino acid residues, whereas typical N-linked glycans contain fewer than 20 sugar residues. For most oligosaccharides, the energy well corresponding to the optimally folded conformation is not nearly as deep as for a protein module and is more like that of a short peptide. As a result, oligosaccharides with identical covalent structures can have different conformations (Figure 7.10). Much of this conformational diversity results from small variations in the torsion angles, but sometimes there are significant differences in the conformations due to large differences in one or more torsion angles. These results demonstrate that the conformation of a glycan is not

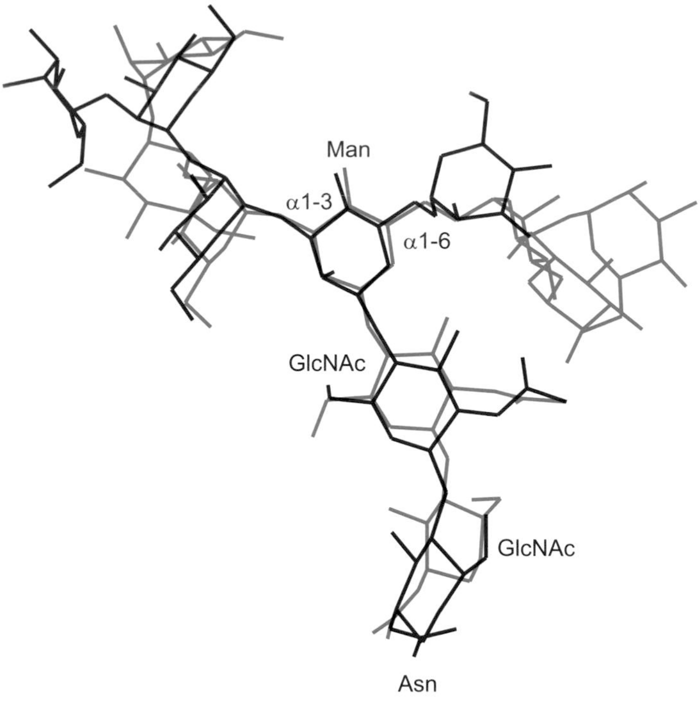

Figure 7.10 Conformations of high mannose N-linked glycans derived from X-ray diffraction of glycoprotein crystals. The overlays illustrate both small variations in torsion angles as well as some larger conformational differences. [Adapted from: Petrescu, A.J., Petrescu, S.M., Dwek, R.A., and Wormald, M.R. (1999). A statistical analysis of N- and O-glycan linkage conformations from crystallographic data, *Glycobiology* 9, 343–352.]

uniquely determined by its covalent structure. The preferred conformations of many glycans may be influenced by interactions with the surfaces of the proteins to which they are attached.

7.7   Oligosaccharide conformations are dynamic

As an additional consequence of the relatively weak energetic stabilization of the preferred conformation of most glycans, the structures are dynamic. The dynamics can be examined by molecular mechanics calculations in which the forces between all the individual atoms in a glycan are calculated allowing for thermal motions of the atoms. Under these circumstances, glycosidic linkages can rotate, passing over energy barriers on the contour maps and assuming different conformations. Some glycans assume a final, most stable conformation, but others continue to flip between two possible sets of torsion angles, suggesting that the two conformations represented by these torsion angles differ little in energy and are not separated by a high energy barrier. Both conformations could exist in an equilibrium mixture. There is experimental evidence that glycans are dynamic, so it may be better to think of them as existing as collections or ensembles of related conformations.

7.8   Short- and long-range interactions also determine the conformations of polysaccharides

The same principles that govern the conformations of oligosaccharides also apply to larger polysaccharides. The preferred torsion angles are determined by interactions between adjacent residues and by residues more distant in the polysaccharide chain. In longer glycan chains consisting of repeated sugar residues in identical linkages, specific combinations of ϕ and ψ angles are repeated over and over again. The cumulative effects of different preferred torsion angles are illustrated by the α1-4-linked glucose homopolymer amylose (Figure 7.11). The α linkage in amylose leads to a bent conformation that is stabilized by hydrogen bonding between adjacent residues. When the resulting torsion angles are repeated at each glycosidic linkage, the extended chain forms a helix.

Similar considerations apply to other homopolymers discussed in other chapters. For example, mannans which form part of the outer wall of many micro-organisms are composed of repeated polymers of mannose in α1-2 linkage. Local interactions between adjacent residues determine which torsion angles are preferred and repetition of these angles at each linkage leads to helix formation (Figure 7.12). The helix may be stabilized by additional hydrogen bonds between non-adjacent residues. Another homopolymer found on bacterial and mammalian cell surfaces is polysialic acid. Electrostatic repulsion between carboxyl groups contributes to the extended helical conformation of these chains. The conformations of simple copolymers containing two alternating types of residues, such as the glycosaminoglycans found in proteoglycans, probably obey similar principles.

> See section 12.7 for more on polysialic acid.

> See section 3.5 for more on glycosaminoglycans.

7.9   Conformations of polysaccharides define properties of cell walls

The predominant polysaccharide in cell walls is cellulose, a repeating polymer of β1-4-linked glucose residues. While the relatively bent α linkages in the polymers discussed in the preceding section favour formation of helical structures, β linkage leads to a more extended structure (Figure 7.12). The preferred torsion angles result from steric considerations plus the formation of hydrogen bonds between adjacent residues. Each polysaccharide chain forms an extended flat rod, with every other glucose residue rotated 180° with respect to its neighbours. The inter-residue hydrogen bonds make the structure rigid. A crystalline cellulose microfibril is generated from the packing together of the polysaccharide strands, first by formation of strand-to-strand hydrogen bonds to form sheets and then by the

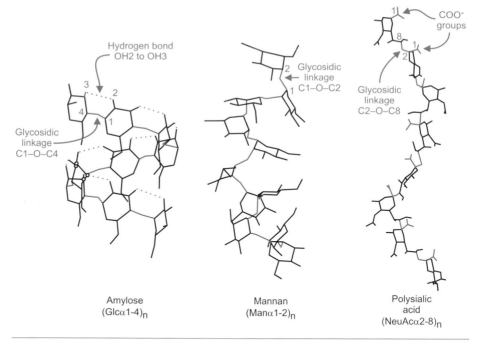

Figure 7.11 Helix formation in linear homopolymeric polysaccharides. Amylose is a storage polysaccharide of glucose found in plants. Hydrogen bonds between adjacent sugar residues favour one conformation. Repetition of the same linkage results in formation of a helix. The preferred conformations of linkages in other polymers are different, resulting in helices with different pitches. Examples include the mannan structure found in the outer wall of the yeast *Candida albicans*, and polysialic acid, which is found on the cell surfaces of mammals and micro-organisms. [Amylose structure adapted from: Winterburn, P.J., Polysaccharide structure and function, in A.T. Bull, J. R. Lagnado, J. O. Thomas, and K.F. Tipton, eds, *Companion to Biochemistry: Selected Topics for Further Study* (London: Longman, 1974), 307–341. Mannan structure adapted from: Nitz, M., Ling, C.-C., Otter, A., Cutler, J.E., and Bundle, D.R. (2002). The unique solution structure and immunochemistry of the Candida albicans β-1,2-mannopyran cell wall antigens, *Journal of Biological Chemistry* **277**, 3440–3446. Polysialic acid structure adapted from: Evans, S.V., Sigurskjol, B.W., Jennings, H.J., Brisson, J.-R., To, R., Tse, W.C., Altman, E., Frosch, M., Weisberger, C., Kratzin, H.D., Klebert, S., Vaesen, M., Bitter-Suermann, D., Rose, D.R., Young, N.M., and Bundle, D.R. (1995). Evidence for the extended helical nature of polysaccharide epitopes: the 2.8 Å resolution structure and thermodynamics of ligand binding of an antigen binding fragment specific for β-(2-8)-polysialic acid, *Biochemistry* **34**, 6737–6744.]

stacking of these sheets. Multiple different forms of cellulose result from slight differences in the packing arrangements, but in each form the stacking of the planes is stabilized by van der Waals packing. Formation of the cellulose microfibrils by alignment of extended strands into stacked sheets is analogous to formation of silk fibres from extended polypeptides in stacks of β sheets and illustrates how macroscopic properties of cellulose emerge from local steric constraints on polysaccharide structure.

Alignment of microfibrils into fibres in a cell wall involves additional types of polysaccharides, including hemicelluloses. A predominant form of hemicellulose is

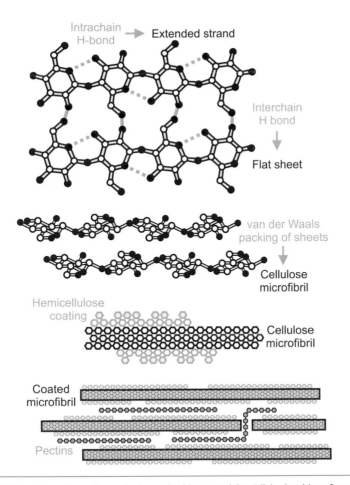

Figure 7.12 Extended conformations of polysaccharides containing β-linked residues found in plant cell walls. The preferred conformation of the β linkage between glucose residues and intrachain and interchain hydrogen bonds stabilize sheets that stack through van der Waals interactions to form cellulose microfibres. Interactions with other polymers, such as hemicellulose and pectins, hold the microfibres in position. [Adapted from: Zugenmaier, P. (2001). Conformation and packing of various crystalline cellulose fibers, *Progress in Polymer Science* **26**, 1341–1417.]

a xyloglucan which, like cellulose, contains a β1-4-linked glucose polymer, but in this case many of the glucose residues are substituted with xylose on the 6 position. The substitution prevents packing of the hemicellulose into fibres, but portions of the backbone polymer can still interact with and coat the surface of the cellulose microfibrils. The less regular structure of the hemicellulose allows it to bridge between the cellulose microfibrils and organize them in three dimensions. Aligned microfibrils are very effective in resisting pulling in the length-wise direction, because all the covalent cellulose polymers run in this direction. Layering of multiple fibres in criss-crossed arrays in the cell wall provides tensile strength in multiple dimensions. Additional acidic polysaccharides, known as pectins, fill spaces between the microfibrils and may be involved in holding them together. These polymers

⊘ See Section 11.3 for more
on signalling in plants.

resemble glycosaminoglycans of animal tissues and may contribute to the hydration properties of the wall. Fragments from the pectins also serve as signalling molecules.

7.10   The conformations of a small number of oligosaccharides have been analysed by X-ray crystallography and nuclear magnetic resonance

The primary experimental tools available for analysis of glycan conformations are X-ray crystallography and NMR spectroscopy. Very few oligosaccharides have been crystallized in isolation and only a modest number of crystal structures have been analysed for glycans attached to proteins. One reason that the conformations of few glycans have emerged from crystallography is that the glycans are often removed from glycoproteins in an effort to eliminate heterogeneity. Unfortunately, this strategy eliminates the possibility of learning about the conformations of the oligosaccharide portions of the glycoproteins. When intact glycoproteins are present in crystals, heterogeneity in the covalent structures of the attached glycans and variability in their conformations often lead to poorly defined electron density, particularly for portions of glycans located furthest from the sites of attachment to the proteins.

The glycans observed in crystal structures can be involved in intermolecular interactions that help to stabilize the crystal lattice. However, the general similarity of the conformations of identical glycans in different crystal environments provides reassurance that these interactions do not significantly perturb the glycan conformations. The number of glycoproteins analysed by crystallography has increased in proportion to the overall exponential increase in the number of protein crystal structures being analysed. Nevertheless, conformations of fewer than 100 N-linked glycans containing five or more residues have been obtained from crystallography.

As in protein structure analysis, NMR is the primary alternative to X-ray crystallography for conformational analysis of glycans. NMR has the potential to yield complementary information about glycans in solution. In some ways, NMR is well suited to the analysis of glycan conformations because typical protein-linked oligosaccharides are less than 5 kDa in size and their water solubility is good. However, interpretation of the NMR spectra of oligosaccharides can be difficult. Within each monosaccharide residue, the protons that yield useful information are those attached to carbon atoms. These protons are almost all in similar chemical environments, because most of the carbon atoms are attached to two other carbon atoms and an oxygen atom. Thus, multidimensional experiments and isotopic labelling may be needed to assign specific NMR signals (resonances) to individual protons in oligosaccharides so that a complete structure can be analysed.

Analysis of conformations by NMR depends heavily on the nuclear Overhauser effect, which is used to detect proximity of two protons in space. Because only a few pairs of protons lie within the detectable range of 5 Å or less, the orientation of two

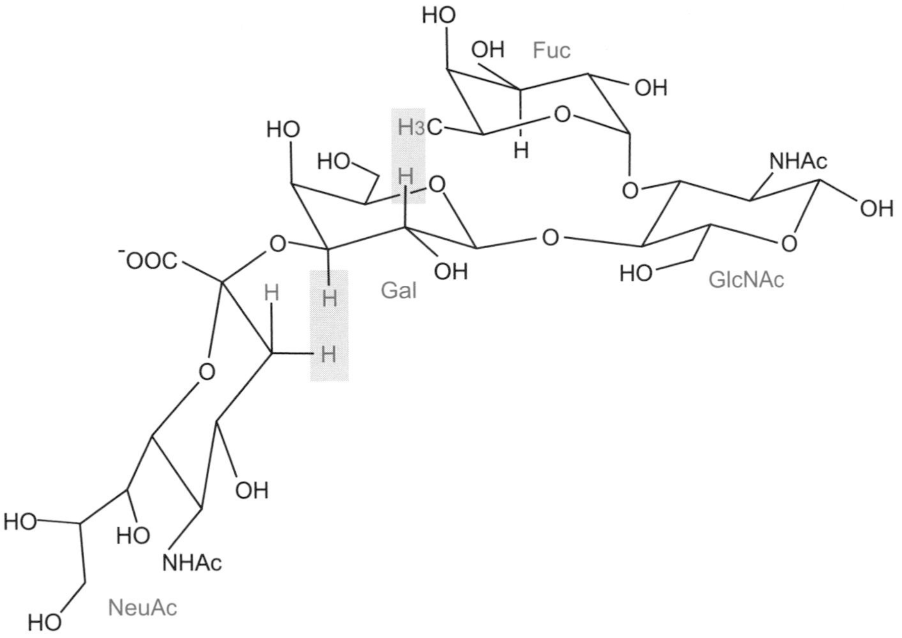

Figure 7.13 Examples of how nuclear Overhauser interactions can be used to establish distances between protons that are used to analyse the geometry of an oligosaccharide. Interactions between the highlighted protons indicate the disposition of the fucose and galactose residues and define the position of the sialic acid residue relative to the galactose residue. [Adapted from: Lin, Y.-C., Hummel, C.W., Huang, D.-H., Ichikawa, Y., Nicolaou, K.C., and Wong, C.-H. (1992). Conformational studies of sialyl Lewis X in aqueous solution, *Journal of the American Chemical Society* 114, 5452–5454.]

sugars relative to each other is often defined by the presence or absence of very few nuclear Overhauser effect signals (Figure 7.13). In spite of these difficulties, NMR has been used to analyse the conformations of about two dozen glycans, sometimes while they are still attached to proteins.

SUMMARY

In spite of the difficulties associated with analysis of the conformations of glycans, we have a reasonable understanding of the possible conformations of the most common types of N-linked glycans. Although the same glycan can assume different conformations under different conditions, the overall shape does not vary dramatically. This structural information forms a basis for understanding the biological roles of glycans discussed in later chapters. Increasing the database of known conformations, not just for the N-linked glycans but for other types of protein- and lipid-linked oligosaccharides, remains a major target for the future.

KEY REFERENCES

Imberty, A. (1997). Oligosaccharide structures: theory versus experiment, *Current Opinion in Structural Biology* **7**, 617–623. This paper gives an overview of the methods available for determining oligosaccharide conformation with reference to determination of the conformations of several biologically interesting glycans.

Nitz, M., Ling, C.-C., Otter, A., Cutler, J.E., and Bundle, D.R. (2002). The unique solution structure and immunochemistry of the *Candida albicans* β-1,2-mannopyran cell wall antigens, *Journal of Biological Chemistry* **277**, 3440–3446. This paper presents NMR experiments demonstrating that antigenic polysaccharides from yeast form helical structures, providing a good example of the use of the nuclear Overhauser effect in determining glycan conformation.

Petrescu, A.J., Butters, T.D., Reinkensmeier, G., Petrescu, S., Platt, F.M., Dwek, R.A., and Wormald, M.R. (1997). The solution NMR structure of glucosylated N-glycans involved in the early stages of glycoprotein biosynthesis and folding, *EMBO Journal* **16**, 4302–4310. This paper provides an example of determination of the conformation of a large, biologically important, oligosaccharide by NMR.

Petrescu, A.J., Petrescu, S.M., Dwek, R.A., and Wormald, M.R. (1999). A statistical analysis of N- and O-glycan linkage conformations from crystallographic data, *Glycobiology* **9**, 343–352. A comparison of all glycan conformations obtained by crystallography allows general conclusions about preferred conformations of particular linkages to be made.

Woods, R.J. (1998). Computational carbohydrate chemistry: what theoretical methods can tell us, *Glycoconjugate Journal* **15**, 209–216. The use of molecular dynamic simulations in determining glycan conformation is reviewed.

Wyss, D.F., Choi, J.S., Li, J., Knoppers, M.H., Willis, K.J., Arulanandam, A.R.N., Smolyar, A., Reinherz, E.L., and Wagner, G. (1995). Conformation and function of the N-linked glycan in the cell adhesion domain of human CD2, *Science* **269**, 1273–1278. This paper presents another example of determination of the conformation of an N-linked oligosaccharide by NMR, in this case with the oligosaccharide still attached to the protein.

QUESTIONS

7.1   On the diagram of the Lewis[x] oligosaccharide shown in Figure 7.9, number the carbon atoms and use this information to deduce which are the B faces of the fucose and galactose residues, and thus to determine which faces interact to stabilize the structure.

7.2   Using molecular models, construct a Glcβ-4Glc disaccharide. Arrange the ϕ angle to the preferred angle resulting from the exo-anomeric effect and then use the fact that there is a hydrogen bond between hydroxyl group 3 of one residue and the ring oxygen of the other to arrange the molecule in the extended β conformation characteristic of cellulose. Estimate the ψ angle in this conformation.

7.3   Compare the molecular organization of polysaccharide in cellulose microfibrils to protein in silk, indicating the conformations of the polymers and the types of bonds that hold the fibres together in each dimension.

7.4 The mannose polymer in the wall of the yeast *Candida albicans* described in Figure 7.11 adopts an unusually compact structure compared with many other sugar polymers. Discuss how the distance measurements derived from NMR experiments were used to determine the conformation, and the implications of the structure for vaccine development.

7.5 The paper by Petrescu *et al.* (1997) in the reference list provides an example of determination of the conformation of a large, biologically important, oligosaccharide by NMR. What information does this study give about the conformation of each linkage in the oligosaccharide? What biological insights can be gained from this study?

Part 2
Glycans in biology

Effects of glycosylation on protein structure and function

LEARNING OBJECTIVES

By the end of this chapter, students should understand:

1 Experimental approaches to studying the effects of glycans on protein structure and function

2 How glycans affect protein folding and stability

3 The effects of glycans on interactions of proteins

In many instances, glycosylation directly affects the properties of proteins and membranes, sometimes with important biological consequences. The examples in this chapter illustrate some of the ways in which glycoprotein functions are modulated by glycosylation, usually by affecting their stability or surface properties. In the first part of the chapter, some experimental approaches to studying the effects of glycosylation on protein structure and function are summarized.

8.1   Various approaches can be used to study the effects of glycosylation

The functions of glycans attached to proteins can be investigated by altering the glycosylation and determining what effect this has. Such experiments can be undertaken at various levels of resolution: elimination of all sugars from all the attachment sites on the protein, elimination of sugars at specific attachment sites, or removal of portions of the sugar structures. N-linked sugars can be completely removed from glycoproteins by treatment with trifluoromethanesulphonic acid or anhydrous hydrogen fluoride. Aside from the difficulty of ensuring that sugar release is quantitative, the major problem with this approach is damage to the protein in the form of occasional cleavage of the polypeptide chain and overall denaturation of the folded structure under harsh conditions. In such circumstances, loss of biological function resulting from such treatments cannot confidently be ascribed to the removal of sugar.

Fortunately, other methods are now available for elimination of N-linked sugars (Figure 8.1). Peptide N-glycanase (PNGase) or one of several endoglycosidases often remove sugars under non-denaturing conditions, although some glycoproteins contain glycans that are inaccessible to these enzymes unless the folded structure is disrupted. When proteins are isolated from cultured cells, it is sometimes possible to use inhibitors of glycosylation such as tunicamycin, which inhibits the initial step in the formation of the dolichol-linked precursor oligosaccharide, to prevent glycosylation in the first place. Alternatively, proteins can be expressed in cells lacking key glycosyltransferases or glycosidases. In a few cases, proteins isolated from tissues consist of mixtures of glycosylated and unglycosylated forms that can be separated and studied.

The methods described above simultaneously probe the role of all the glycans attached to a protein at the same time. Additional methods can be used to test the effects of glycan attachment at specific sites. Site-directed mutagenesis to alter glycosylation target sequences in proteins makes it possible to study the effects of individual glycans. For example, N-linked glycosylation at a particular site can be eliminated by changing either the asparagine attachment residue, or the serine or threonine side chain two residues away.

The roles of individual sugars in glycan chains can be tested with exoglycosidases. This approach has been particularly useful for demonstrating the importance of terminal sialic acid residues, because sialidases (neuraminidases) usually work on glycans that are still attached to proteins. Unfortunately, few of the other glycosidases are as accessible and many do not work efficiently on glycans covalently linked to proteins. Because many proteins are heterogeneously glycosylated, it should theoretically be possible to separate and compare the activities of different glycoforms. However, it

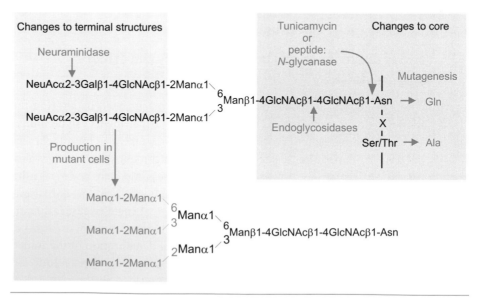

Figure 8.1 Approaches to preparation of glycoproteins that lack some or all of their N-linked oligosaccharides.

is usually difficult to perform such separations on a large enough scale to make such experiments practical. Where proteins can be expressed in transfected tissue culture cells, mutant cell lines that lack specific glycosyltransferases and glycan-processing enzymes can be used to produce proteins with altered sugars. For example, the commonly used Lec1 line of Chinese hamster ovary cells produces glycoproteins that contain only high mannose N-linked oligosaccharides instead of complex glycans.

> ➲ See **Section 5.7** for more on mutant cell lines.

While there are many techniques for testing hypotheses about the significance of individual sugar residues in glycoproteins, these approaches are generally not as specific as the site-directed mutagenesis of polypeptides. The methods must be applied artfully and interpreted with great caution. As a result of these considerations, there are relatively few instances in which we have a detailed molecular understanding of how a particular glycan confers a specific biological property on a glycoprotein.

8.2 Sugars stabilize the structure of the cell adhesion molecule CD2

CD2 is a cell adhesion molecule on the surface of killer T lymphocytes that mediates binding to target cells. The extracellular portion of the CD2 polypeptide consists of two domains from the immunoglobulin superfamily that are linked to the membrane by a C-terminal membrane anchor (Figure 8.2). Counter-receptors for CD2 on target cells, CD58 in humans and CD48 in mice, have similar overall domain organization and cell adhesion results from interactions between the N-terminal immunoglobulin-type domains.

There are three N-linked glycosylation sites in CD2, one in each of the immunoglobulin-type domains and one in the interdomain linker sequence. Several techniques have been used to demonstrate that glycosylation of human CD2 is essential for its biological function. Treatment of CD2 with PNGase to remove all N-linked carbohydrate results in complete loss of binding to CD58. Mutation of the glycosylation site in the N-terminal domain, either by changing the glycosylated Asn65 to glutamine or by changing Thr67 to alanine also eliminates binding activity, thus indicating an essential role for the N-linked carbohydrate at position 65. Structural analysis shows that this site is occupied by high mannose oligosaccharides bearing 5–9 mannose units. From these results, it can be concluded that this oligosaccharide is essential for the function of CD2.

The N-terminal immunoglobulin-type domain of CD2 consists of a sandwich of two β-sheets. The surface that interacts with CD58 lies on the outside of one face of the sandwich. Surprisingly, the glycosylation site lies almost precisely on the opposite side of the domain, on the outside surface of the other β-sheet. Thus the attached carbohydrate does not form part of the binding site. Insight into the effect of the carbohydrate comes from comparison of rat and human CD2. Rat CD2 is not glycosylated at the position corresponding to Asn65 in human CD2, but there is an additional important difference between the human and rat proteins. In the human protein, lysine residues at positions 61, 69, and 71 all project from the same side of the molecule as the N-linked carbohydrate. These lysine residues form a cluster of

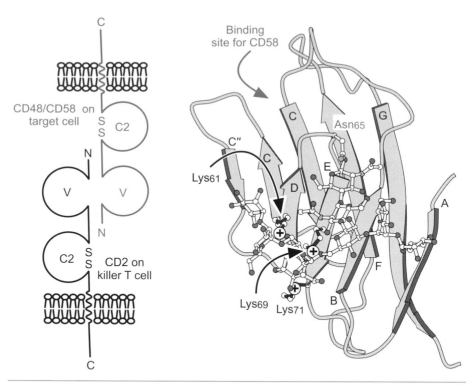

Figure 8.2 CD2 structure and interaction with counter-receptors CD48/CD58. Each extracellular domain consists of two immunoglobulin superfamily domains. The binding site of CD2 is located on the back face of the back β-sheet consisting of strands C, C′, and F plus the connecting loops. The essential glycosylation site is located on the front β-sheet at Asn65. [Based on entry 1GYA in the Protein Databank.]

positive charges. In the rat homologue, this cluster is interrupted by a glutamic acid residue in position 61. It appears that the role of the carbohydrate in human CD2 may be to stabilize the cluster of positive charges by making hydrogen bonds with the amino groups of the lysine residues. No such stabilization would be required in the case of rat CD2 because of the negatively charged side chain at position 61. Human CD2 in which lysine at position 61 has been replaced with glutamic acid is functional even in the absence of glycosylation at position 65.

Only the first three residues of the glycan are required to stabilize CD2, which is consistent with the idea that the inner-most mannose residue can interact with the lysine residues and stabilize the folded state (Figure 8.2). However, some stabilization is also observed with only the first GlcNAc residue attached to the protein, which cannot be explained in this way. Stabilization resulting primarily from attachment of the first GlcNAc residue to asparagine is commonly observed for other glycoproteins. Although this effect could result from local interactions with the polypeptide backbone, computational modelling of glycoprotein stability suggests the alternative possibility that the addition of the sugar residues favours the folded form of the protein by destabilizing the unfolded state rather than stabilizing the folded state (Figure 8.3).

The basis for the destabilization is unclear, but it may be that the sugar groups prevent transient local interactions that normally stabilize the unfolded polypeptide.

Several important conclusions can be drawn from these experiments. The studies indicate that an essential role for carbohydrate in a biological function of a glycoprotein does not necessarily imply a direct role in mediating that function. In this instance, glycosylation is needed to stabilize the protein structure but does not directly participate in the interaction with a counter-receptor. In addition, comparison of the rat and human proteins indicates that, although the carbohydrate can stabilize the protein structure, this is not the only way such stabilization can be achieved. Changes elsewhere in the amino acid sequence, such as the glutamic acid for lysine substitution at position 61, result in a stable, unglycosylated structure.

8.3   An oligosaccharide replaces an α-helix in some variant surface glycoproteins of trypanosomes

Rat and human CD2 represent two solutions to a structural problem, one achieved by glycosylation and the other by an amino acid change. Among examples of this phenomenon, one of the best characterized occurs in the family of variant surface glycoproteins that coat the surface of the trypanosome parasite (Figure 8.4). Trypanosomes are large, single-celled eukaryotes that move between multiple hosts and cause sleeping sickness in humans. During an important phase of their life cycle, trypanosomes move about in the circulation. In order to survive, they evade host immune responses by sequentially expressing a series of different coat proteins, the variant surface glycoproteins, which are encoded by a large family of genes. These

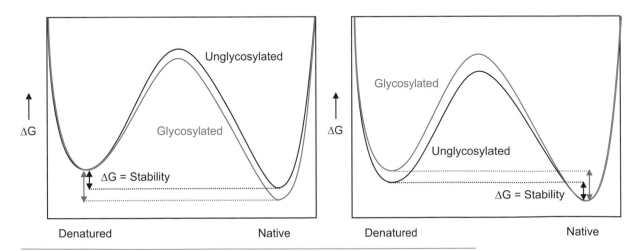

Figure 8.3 Potential mechanisms of enhancing protein stability through glycosylation. In the situation shown on the left, contacts between the glycan and the natively folded protein stabilize the folded conformation. Alternatively, as shown on the right, glycosylation can destabilize the unfolded state, making the folded state more favourable.

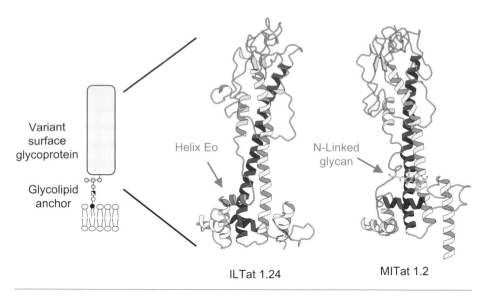

Figure 8.4 Variant surface glycoproteins that form a surface coat outside the plasma membrane of trypanosomes. The structures of two variants are similar in overall fold but differ in details, including the presence of an N-linked glycan in place of helix E_o covering a hydrophobic patch on the stalk structure. [Based on entries 1VSG and 2VSG in the Protein Databank.]

genes have been under strong evolutionary pressure to diverge sufficiently so that antibodies and killer T cells reactive with one of the protein variants do not react with another. As the host antibodies and T cells begin to recognize the trypanosomes, the parasite switches to expression of a different variant. Thus, the parasite always stays one step ahead of the immune response of the host, which results in the characteristic waves of fever associated with trypanosomal infections.

The structures of the extracellular portions of two trypanosomal variant surface glycoproteins have been solved by X-ray crystallography. Although these two proteins show only 16% sequence identity, approximately 60% of the structural elements are the same, confirming that they have arisen by a process of divergent evolution from a common ancestor. At the base of the structure, the MITat 1.2 variant has an asparagine-linked high mannose oligosaccharide that is absent from the ILTat 1.24 variant because the sequence has diverged and the necessary Asn-X-Ser/Thr sequence is not present. An extra short stretch of α-helix (helix E_o) in the ILTat 1.24 variant covers a hydrophobic patch in the region where the B and H helices cross. In the MITat 1.2 variant, the high mannose oligosaccharide shields this patch. Thus, segments of polypeptide and oligosaccharides can achieve independent solutions to a common structural problem.

In the CD2 and trypanosome examples, it may seem perplexing that glycosylation has been used as a means of stabilizing a particular structure when simple changes to amino acid sequences can achieve the same ends. Such changes would not require the complex machinery for synthesizing oligosaccharides and attaching them to proteins. The key to understanding this apparent paradox is to remember that the glycosylation machinery probably did not evolve to solve this type of structural problem. It arose to carry out the other functions introduced in Chapter 1 and discussed in more detail in the rest of this book. However, once established, the machinery will work on all proteins that pass through the endoplasmic reticulum on their way to the cell surface. If, as a result of random mutations, an Asn-X-Ser/Thr sequence appears on the surface of a protein destined for the plasma membrane, it will be glycosylated. Evolutionary selection will then act on the resulting glycosylated structure. If the glycosylated asparagine by chance stabilizes the protein, selection will favour its retention.

In the case of the trypanosome surface glycoproteins, a glycosylation site may have appeared in an ancestor that looked like the ILTat 1.24 variant and would initially have been neutral in terms of evolutionary fitness. However, if further mutations resulted in loss of the E_o helix, the presence of the oligosaccharide would now be advantageous and would be subject to positive selective pressure to ensure that it is retained to cover the hydrophobic patch. Similar arguments can be made for CD2. Thus, the oligosaccharides in these molecules do not perform unique functions that could not be achieved with amino acids alone. In the context of the existing glycosylation machinery, essential for other purposes, an asparagine in the Asn-X-Ser/Thr context becomes a novel oligosaccharide-conjugated amino acid that is then selected for (or against) as a whole. This extra structural unit is used in a variety of ways during evolution.

8.4   Attachment of a monosaccharide can affect protein dynamics

Glycosylation can modulate the stability and the dynamics of proteins in subtle ways. Understanding the basis for such effects requires a rather detailed knowledge of the structure of the glycoprotein and its unglycosylated counterpart, so small proteins such as protease inhibitors are often favoured for model studies. For example, the PMP-C protease inhibitor isolated from locusts is a 36-amino acid polypeptide that contains a single O-linked fucose residue attached to threonine at position 9. The effect of glycosylating the PMP-C inhibitor has been examined by comparing the structures of two chemically synthesized versions of the polypeptide, one with fucose attached to Thr9 and one lacking it. The structure, determined by NMR spectroscopy, consists of a three-stranded β-sheet that is essentially unaffected by the presence of the fucose residue (Figure 8.5). Comparison of the chemical shifts of individual atoms, which are sensitive to the local environment of these atoms, reveals that the only conformational differences between the glycosylated and unglycosylated versions of the inhibitor are within 6 Å of the fucose residue, including the adjacent amino acid residues Thr8 and Phe10.

The functional consequences of these local interactions can be seen in studies of the stability of the inhibitor as a function of temperature. The unfucosylated form of the protein denatures at a temperature approximately 20°C lower than the fucosylated form, demonstrating that interactions of the fucose contribute to the stability of the folded form of the protein. The long-range effects of local interactions with the sugar are demonstrated by measurements of deuterium exchange of amide protons in the polypeptide backbone,

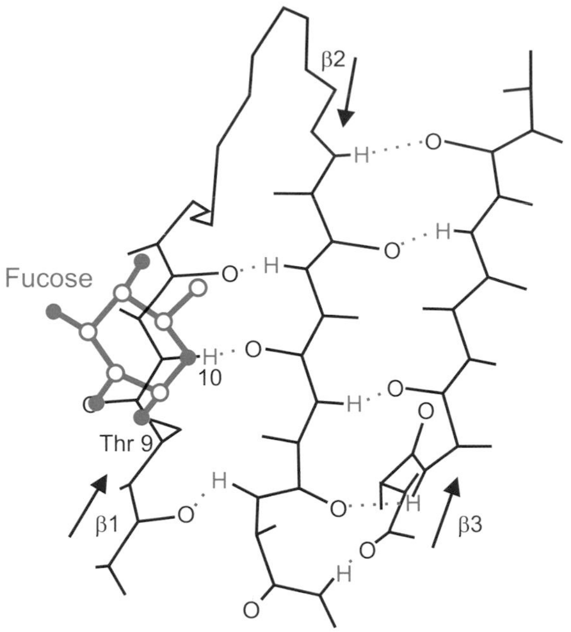

Figure 8.5 Structure of the β-sheet core of the PMP-C protease inhibitor. Formation of the highlighted hydrogen bonds between β strands reduces the rate of exchange of these protons. The threonine residue stabilizes the entire β sheet and thus reduces proton exchange even at residues that are distant in the structure.

which reflect the degree to which these atoms are exposed to solvent. Residues involved in hydrogen bonding between strands of the β-sheet show low rates of exchange with deuterated water. At room temperature, the exchange rates for all of these amide protons are lower in the fucosylated than in the unfucosylated protein.

Thus the protein exists as an equilibrium mixture of two states: a fully folded form in which the β-sheet is present and a fully unfolded form in which the β-sheet has come apart. This interpretation emphasizes the global and co-operative nature of protein folding. Fucose shifts the equilibrium toward the folded state even though it is located only at one edge of the β-sheet, while in the absence of the fucose residue the entire structure spends more time in the unfolded form. Thus, fucose stabilizes hydrogen bonds on the far side of the β-sheet by making local interactions that determine the equilibrium state of the entire module.

8.5 Glycosylation affects the ability of immunoglobulins to fix complement and bind to receptors

A primary function of antibodies is to target killing of pathogens by interacting with serum complement proteins that directly lyse cells and by binding to Fc receptors that lead to uptake by macrophages. In the case of IgG, both complement fixation and binding to Fc receptors require interactions with the two C-terminal constant domains of the heavy chain that form the Fc portion of the immunoglobulin. A conserved glycosylation site is found in this portion of the molecule and many different types of experiment have demonstrated that a glycan at this site is required for IgG to participate in complement fixation and receptor binding. However, structural analysis reveals that there is little or no direct contact between the glycan and the complement components and receptors that bind to the IgG. The glycan, which is attached to the C_H3 domain, has an unusual location inside rather than on the surface of the protein and it makes extensive contacts with the C_H3 and C_H2 domains (Figure 8.6). In the absence of the glycan, these domains move toward each other into a conformation that cannot interact with the Fc receptor.

IgG provides an example of a glycoprotein in which the glycosylation plays an essential structural role so that its absence has direct functional consequences. The requirement for glycosylation would appear to preclude production of functional antibodies in bacteria. However, following appropriate selection, it is possible to identify mutant forms of the heavy chain of IgG that retain the ability to bind the Fc receptor even in the absence of glycosylation. The structural effects of these changes are not known, but this finding appears to reflect the principle noted with CD2 and the variant surface glycoproteins of trypanosomes, since a similar structural feature can be achieved using either glycan side chains or amino acids in the polypeptide.

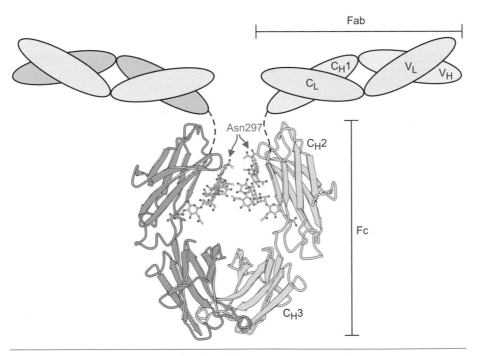

Figure 8.6 Location of N-linked glycans in IgG. The conserved IgG glycan fits into a pocket in the Fc region, resulting in sequestration of the glycan and preventing the C_H2 domains from approaching each other. [Based on entry 1FC1 in the Protein Databank.]

8.6   Protein–protein interactions can be modulated by oligosaccharides

Sugars on the surfaces of proteins would not be expected to have a significant effect on the ability of small molecule substrates and ligands to reach active sites and other binding pockets. However, relatively bulky glycan substituents can reduce the accessibility of the proteins to larger probes such as antibodies, proteases, and other proteins. Shielding of hydrophobic regions of a protein surface from the aqueous solvent can also increase solubility. This phenomenon has often been noted in the case of recombinant proteins produced in bacterial systems. These proteins lack both N- and O-linked glycosylation, and they are often less soluble than their glycosylated counterparts produced in eukaryotic expression systems.

Ways in which exposure of extra surfaces by deglycosylation or underglycosylation of protein can lead to unexpected behaviours of proteins are well illustrated by studies of glycoprotein hormones produced in the pituitary gland and the placenta. This family of related hormones includes lutropin, follitropin, and chorionic gonadotropin. The primary role of some of the N-linked glycans on these hormones is probably to direct their clearance from circulation, which is mediated by a receptor in liver. The receptors for these hormones on target cells in the gonads are distinct from the clearance receptors and are members of the seven transmembrane helix family. Binding of the intact, glycosylated hormones to target cell receptors leads to the activation of adenylate cyclase, but removal of the N-linked glycans abolishes their ability to activate these receptors. This inactivation is seen when the entire glycan is absent, either following deglycosylation or when key glycosylation sites are removed by site-directed mutagenesis. However, removal of just the terminal sugars does not affect hormone activity even though this treatment can affect interaction with the clearance receptors.

These results might suggest that the core of the N-linked oligosaccharides plays a direct role in hormone binding to target cell receptors. However, the affinity of the deglycosylated hormones for their target cell receptors actually increases even though the bound hormones do not cause receptor activation. The most likely explanation for this phenomenon is that the absence of the glycans probably exposes new surfaces that interact in abnormal ways with the receptor (Figure 8.7). These binding interactions result in increased affinity but also shift the hormone out of the position needed for receptor activation. These results illustrate the degree of caution that must be exercised in interpreting the results of deglycosylation experiments.

The fact that glycans can have multiple roles in the function of serum glycoproteins can make it difficult to interpret deglycosylation experiments. The importance of N-linked glycans attached to the hormone erythropoietin has been extensively investigated because this hormone is widely used in the treatment of anaemia. There have been conflicting claims about the importance of the

⊖ For more on glycoprotein hormone clearance see Section 10.9.

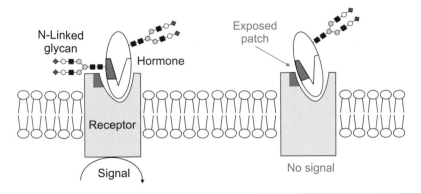

Figure 8.7 Interaction of pituitary and placental glycoprotein hormones with receptors that activate adenylate cyclase. Removal of specific glycosylation sites results in increased affinity for the receptor, possibly because an exposed patch on the surface of the hormone makes novel interactions with the receptor. These interactions do not lead to receptor activation. Although typical bi-antennary structures are shown in the diagram, different glycoprotein hormones bear different glycans (see Chapter 10).

N-linked glycans attached to erythropoietin, probably because they have many different roles. Glycosylation affects the stability of the hormone as well as its ability to activate its receptor. In assays with cells *in vitro*, underglycosylation can lead to increased affinity for the erythropoietin receptor. In this case, the increased affinity is accompanied by increased receptor activation. However, as with other serum glycoproteins, glycosylation is important for the trafficking of newly made erythropoietin out of kidney cells, and it affects the lifetime of the hormone in the circulation. Because lack of glycosylation can lead to rapid clearance from circulation as well as decreased stability, underglycosylated erythropoietin is relatively inactive in an organism despite its increased activity in cell-based assays.

⊘ See **Box 10.2** for more on erythropoietin.

8.7 Oligosaccharides covering surfaces of proteins can protect against proteolysis

If a protein–glycan linkage is flexible rather than fixed, the glycan may at least partially cover a large portion of the protein surface, although the effect of the glycan on accessibility to any given part of the protein surface may well be transient. Such transient blocking explains the fact that glycosylation can sometimes confer increased resistance to proteolysis without preventing it altogether. Such modulation of proteolysis is illustrated by tissue plasminogen activation. The protease plasmin mediates removal of blood clots by degrading fibrin when tissue integrity has been restored and the clot has served its purpose in preventing blood loss (Figure 8.8). Active plasmin is generated by **tissue plasminogen activator**, which

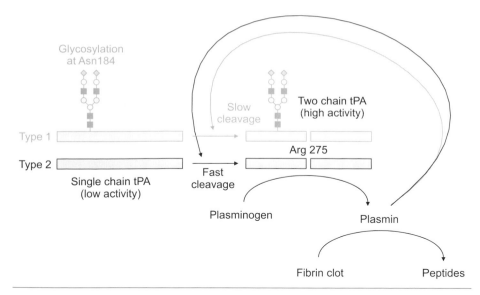

Figure 8.8 Feedback loop for plasmin cleavage of tissue plasminogen activator (tPA). The two-chain form of tissue plasminogen activator is at least five-fold more active than the single chain form. Both types of tPA bear N-linked glycans at positions 117 and 448. They differ at position 184, which is glycosylated in type 1 tPA and unglycosylated in type 2 tPA. The absence of glycosylation at position 184 allows more rapid proteolytic activation of type 2 tPA.

cleaves the precursor plasminogen. Tissue plasminogen activator exists in a single chain and a two-chain form. Cleavage of the single chain form to the more active two-chain form is catalysed by plasmin, thus forming a positive feedback loop. Type 1 tissue plasminogen activator, which is glycosylated at residue Asn184, is relatively resistant to proteolytic activation compared with type 2, which is not glycosylated at this position. Thus the presence of a glycan attached to Asn184 impedes access of plasmin to the Arg275 cleavage site needed to generate the more active two-chain tissue plasminogen activator. It might be useful to have multiple forms of tissue plasminogen activator that can be activated at different rates. However, if the activation rate of tissue plasminogen activator with or without a glycan at Asn184 is sufficient to control clotting, the variability could simply result from the vagaries of the glycosylation machinery.

The physical covering of protein surfaces may sometimes provide a general resistance to attack by proteases. It is certainly true that proteins are often more susceptible to protease digestion following deglycosylation, although it is sometimes difficult to know if selection for this property accounts for the degree of glycosylation of such proteins. It seems unlikely that specific sugar structures would be selected to serve this function because a range of glycans would serve the same purpose. As in the case of neutral evolution of protein structure, divergence of glycan structures would be tolerated.

SUMMARY

A wide variety of tools are available for producing glycoproteins with reduced or altered glycosylation. In many cases, changes to glycosylation are neutral in terms of their effect on protein function, but attached glycans can affect the structure and activity of proteins. While the attachment of sugars usually does not alter a stable, folded protein conformation, sugars can stabilize this conformation by adding hydrogen bonding and hydrophobic interactions. Because sugars are not uniquely able to cause this type of stabilization, it is unlikely that this function of glycosylation was the driving force for evolution of the glycosylation machinery. However, with the machinery in place, natural selection has taken advantage of the stabilization that glycosylation can afford. Sugars also serve to shield various portions of protein surfaces from the aqueous solvent and from proteases. In many cases, it is the core structures and the overall bulk of the glycans that provides these functions, so variability in terminal structures is tolerated.

KEY REFERENCES

Blum, M.L., Down, J.A., Gurnett, A.M., Carrington, M., Turner, M.J., and Wiley, D.C. (1993). A structural motif in the variant surface glycoproteins of *Trypanosoma brucei*, *Nature* **362**, 603–609. Crystallographic analysis of the structures of two trypanosome variant surface glycoproteins is presented.

Feige, M.J., Nath, S., Catherino, S.R., Weinfurtner, D., Steinbacher, S., and Buchner, J. (2009). Structure of the murine unglycosylated IgG1 Fc fragment, *Journal of Molecular Biology* **391**, 599–608. Structural changes that underlie the different behaviour of glycosylated and unglycosylated IgG are described.

Mer, G., Hietter, H., and Lefevre, J.-F. (1996). Stabilization of proteins by glycosylation examined by NMR analysis of a fucosylated proteinase inhibitor, *Nature Structural Biology* **3**, 45–53. Experiments showing stabilization of the proteinase inhibitor PMP-C by O-linked fucose are described.

Purohit, S., Shao, K., Balasubramanian, S.V., and Bahl, O.P. (1997). Mutants of human choriogonadotropin lacking N-glycosyl chains in the α-subunit: mechanism for the differential action of the N-linked carbohydrates, *Biochemistry* **36**, 12355–12363. Analysis of the effects of N-linked glycans of choriogonadotropin on structure and activity of the hormone is described.

Shental-Bechor, D. and Levy, Y (2009). Folding of glycoproteins: toward understanding the biophysics of the glycosylation code, *Current Opinion in Structural Biology* **19**, 524–533. A computations analysis of the effects of glycans on the stability of the native and unfolded states of proteins is presented.

Wittwer, A. and Howard, S.C. (1990). Glycosylation at Asn-184 inhibits the conversion of single-chain tissue-type plasminogen activator by plasmin, *Biochemistry* **29**, 4175–4180. Key experiments showing that glycosylation modulates tPA activity by inhibiting proteolysis are presented.

Wormald, M.R. and Dwek, R.A. (1999). Glycoproteins: glycan presentation and protein-fold stability, *Structure* **7**, R155–R160. Several examples of proteins stabilized by the presence of N-linked glycans are reviewed.

Wyss, D.F., Choi, J.S., Li, J., Knoppers, M.H., Willis, K.J., Arulanandam, A.R.N., Smolyar, A., Reinherz, E.L., and Wagner, G. (1995). Conformation and function of the N-linked glycan in the cell adhesion domain of human CD2, *Science* **269**, 1273–1278. Experiments showing that CD2 is stabilized by a high mannose N-linked oligosaccharide are described.

QUESTIONS

8.1   What properties of a glycoprotein might you expect to be changed by a mutation that deleted a glycosylation site?

8.2   You are studying the mechanism of action of an enzyme that contains one sialylated, tri-antennary complex N-linked oligosaccharide. When you express the enzyme in *Escherichia coli* and assay its activity, you find that this recombinant form of the enzyme has only 25% of the activity of the native enzyme. What approaches would you take to investigate the importance of the N-linked oligosaccharide for the activity of the enzyme?

8.3   Discuss the experiments used to show that N-linked oligosaccharides protect some lysosomal membrane proteins from proteolysis.

Reference: Kundra, R. and Kornfeld, S. (1999). Asparagine-linked oligosaccharides protect Lamp-1 and Lamp-2 from intracellular proteolysis, *Journal of Biological Chemistry* **274**, 31039–31046.

8.4   How does glycosylation of the IgG Fc region affect its function?

Reference: Radaev, S. and Sun, P.D. (2001). Recognition of IgG Fcγ receptor: The role of Fc glycosylation and the binding of peptide inhibitors, *Journal of Biological Chemistry* **276**, 16478–16483.

Radaev, S., Motyka, S., Fridman, W.-H., Sautes-Fridman, C., and Sun, P.D. (2001). The structure of a human type III Fcγ receptor in complex with Fc, *Journal of Biological Chemistry* **276**, 16469–16477.

Carbohydrate recognition in cell adhesion and signalling

LEARNING OBJECTIVES

By the end of this chapter, students should understand:

1 The classification of sugar-binding proteins based on structures of carbohydrate-recognition domains

2 The roles of mannose-binding protein and other C-type lectins in innate immunity

3 How the selectins function as cell adhesion molecules

4 The adhesion and signalling functions of siglecs in the immune system

5 How galectins modulate adhesion and signalling in lymphocytes and other cells

Lectins that serve as receptors for specific glycans have been referred to in previous chapters. This chapter and Chapter 10 focus on biological processes that are performed by lectins. Some of the lectins that are discussed recognize foreign cell surfaces and mediate or modulate immune responses to pathogens, while others bind to endogenous carbohydrates and mediate adhesion or signalling events at the cell surface. Before discussing these specific examples, it is useful to begin with a summary of some of the features of lectin molecules.

9.1   Animal lectins can be classified based on their structures

The sugar-binding activities of lectins usually reside in discrete protein modules that are referred to as carbohydrate-recognition domains (CRDs). Classification of lectins is based on the structures of these domains (Table 9.1). The CRDs are responsible for the recognition functions of the lectins, while other domains in the lectin polypeptides mediate subsequent responses to the recognition events. In general, different types of CRDs have distinct ancestry and different polypeptide folds. The overall folds of CRDs are determined by conserved residues that are characteristic of each family of CRDs. These residues form the hydrophobic core and in some cases disulphide bonds and binding sites for divalent cations.

Table 9.1 Types of carbohydrate-recognition domains (CRDs)

Type	Structure	Typical ligands	Examples of functions
Calnexin	β-sandwich	Glc_1Man_8	Protein sorting in the endoplasmic reticulum
M-type	α-helical barrel	Man_8	Endoplasmic reticulum-associated protein degradation
L-type	β-sandwich	Man_{5-9}	Protein sorting in the endoplasmic reticulum
P-type	Unique β-rich structure	Man 6-phosphate	Protein sorting post-Golgi
C-type	Unique mixed α/β structure	Various	Cell adhesion (selectins)
			Glycoprotein clearance
			Innate immunity (collectins)
Galectins	β-sandwich	b-galactosides	Glycan cross-linking at the cell surface
I-type	Immunoglobulin superfamily	Sialic acid	Cell adhesion (siglecs)
R-type	β-trefoil	Various	Enzyme targeting
			Glycoprotein hormone turnover

Although the CRD classification scheme is based on protein sequences, CRDs of a particular type tend to share glycan-binding properties. For example, the C-type CRDs bind sugars in a Ca^{2+}-dependent manner. However, this property is not, by itself, a distinguishing feature of the C-type CRDs because calnexin, calreticulin, and the L-type lectins all show Ca^{2+} dependence of ligand binding. It is also noteworthy that, in some cases, the glycans bound by CRDs of a particular type are closely related in structure, while in other cases they are very diverse. In spite of the fact that different CRD families assume very different protein folds, there are common themes in the way that these diverse proteins bind sugars selectively; these are highlighted in this and the next chapter.

9.2   Mannose-binding protein is a host defence molecule that initiates the lectin pathway of complement activation

Serum mannose-binding protein (MBP), which is also known as mannose-binding lectin (MBL), binds to carbohydrate structures on the surfaces of micro-organisms, using the ability to distinguish such exogenous structures from mammalian glycans to provide a mechanism for identification and neutralization of pathogens. Thus, MBP that is constitutively circulating in the bloodstream provides a type of innate immune response. Innate immunity is distinguished from adaptive immunity, which requires the production of specific antibodies and cytotoxic T cells in response to the presence of a foreign substance.

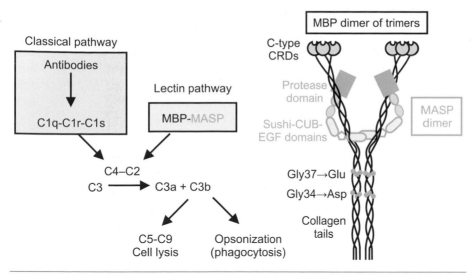

Figure 9.1 Summary of the role of mannose-binding protein (MBP) in innate immunity. The complex of MBP with MBP-associated serine protease (MASP)-2 can activate the C3 convertase enzyme (C4–C2) by a mechanism analogous to the activation of the classical pathway through complement component C1qC1rC1s. Activation requires multivalent interaction of the MBP CRDs with arrays of cell surface glycans. EGF, epidermal growth factor.

MBP oligomers are assembled in two stages (Figure 9.1). First, three polypeptides associate to form a trimer that is stabilized by two helical regions, a neck consisting of a coiled coil of α-helices, and an extended tail that forms a collagen-like helix. Lateral association of these trimeric building blocks leads to assembly of larger oligomers consisting of six or more polypeptides. Molecules with this overall organization of collagenous and lectin domains are called **collectins**.

When the C-terminal recognition portion of MBP binds to carbohydrates on the surfaces of micro-organisms, the N-terminal collagen-like domain can interact directly with a **collectin receptor** on macrophages, leading to phagocytosis. More commonly, the N-terminal collagen-like domains initiate the **lectin branch** of the **complement pathway**. In the **classical pathway** of **complement fixation**, antibody binding to pathogens leads to fixation of the first component of complement, C1q. The C1q molecule has a bouquet-like structure that is held together by collagen-like stalks and appears very similar to MBP. C1q associates with two serine proteases, C1r and C1s, which initiate a proteolytic cascade by activating complement components C4 and C2. MBP activates the same cascade through **MBP-associated serine proteases (MASPs)**. Activated C2–C4 complexes resulting from either pathway cleave complement component C3, resulting in deposition of a large C3b fragment on the target surface. C3b mediates phagocytosis of pathogens by interacting with a receptor on macrophages. In addition to this opsonic activity, immobilized C3b also initiates a lytic pathway in which additional complement components assemble to form a membrane pore. It is not completely understood how binding of MBP to a carbohydrate-rich surface triggers activation of MASPs, but the process involves lateral interactions between the clusters of CRDs on the surface, which alter the angles between the collagen-like stalks in the region where MASPs are bound.

The importance of the innate immune response mediated by MBP is evident from disease conditions associated with loss of MBP activity from serum. Mutations in the collagen-like domain of MBP are associated with susceptibility to recurrent, severe infections in children (particularly between six months and two years of age). This immunodeficiency reflects the gradual loss of maternal immunity provided by antibodies that either crossed the placenta or were taken up from the mother's milk, at a time when the full repertoire of the infant's own immune system is still developing. The role of MBP is also evident in patients with acquired immunodeficiency syndrome (AIDS). Such patients have a compromised ability to mount an adaptive immune response and rely on the innate protection provided by MBP. If they have an inherited MBP deficiency as well, they are more susceptible to opportunistic infections and have a shorter life expectancy.

The molecular basis for MBP deficiency is well understood. The two most common mutations leading to this condition result in changes of glycine residues within the collagen-like domain. Insertion of either glutamic acid or aspartic acid side chains disrupts the normal structure of this domain, because every third residue must be glycine in order to allow packing into the triple helical conformation (Figure 9.1). The disruption destabilizes the surrounding regions of collagen-like structure, including the MASP binding site. Defective MASP binding is the major contributor to the deficiency phenotype. Although the disease susceptibility is most pronounced for individuals who are homozygous for any of these mutations, it is also seen in heterozygous individuals. This largely dominant behaviour is a result of the mixing of wild-type and mutant forms of the MBP polypeptide, so that essentially no oligomers containing completely correctly folded collagenous domains are made.

Two other collectins, lung surfactant proteins A and D (SP-A and SP-D), are important in the innate immune response. These two proteins, which form part of the mixture of phospholipids and proteins lining the alveolar spaces, help to protect against micro-organisms that enter the lung. Like MBP, both SP-A and SP-D bind sugars on the surfaces of pathogens through their C-type CRDs. However, the collagenous domains of the lung collectins do not initiate complement fixation. Instead, the collagenous domains of SP-A and SP-D interact with receptors on alveolar macrophages, leading to phagocytosis of the bound micro-organisms.

9.3   Pathogen recognition by mannose-binding protein results from both monosaccharide-binding specificity and oligomer geometry

Binding of complex ligands to C-type animal lectins is based on primary monosaccharide binding sites that are combined or extended to allow selective recognition of oligosaccharides. The ability of MBP to distinguish between self and non-self is based on the different compositions and structures of mammalian glycoproteins and glycolipids compared with those on the surfaces of most micro-organisms. Glycans on bacteria, yeast, and fungi differ widely but many share a few key characteristics. Because these glycans usually serve structural roles, they tend to have repetitive structures and these structures often present terminal glucose, GlcNAc, and mannose residues (Figure 9.2). Recognition by MBP takes advantage of both of these features. In spite of its name, MBP shows equal affinity for glucose, mannose, and GlcNAc residues in terminal positions.

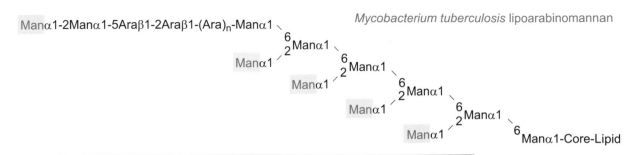

Figure 9.2 Examples of surface glycans found on potentially pathogenic micro-organisms. Terminal residues that can serve as ligands for mannose-binding protein and the mannose receptor are highlighted in blue.

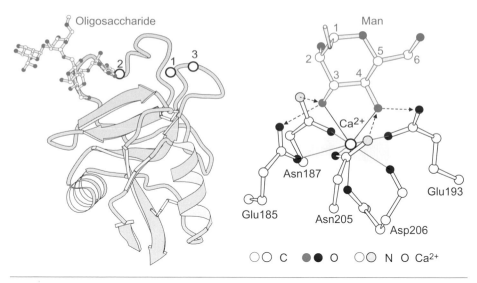

Figure 9.3 Structure of the CRD of mannose-binding protein with bound oligosaccharide. A close-up of the binding site around Ca^2 is shown on the right. Hydrogen bonds are indicated with dashed arrows pointing in the direction from donor to acceptor. Solid lines denote coordination bonds, blue for axial and black for equatorial. [Based on entry 2MSB in the Protein Databank.]

The sugar-binding site in the C-type CRD from serum MBP is located adjacent to one of the Ca^{2+} (Figure 9.3). The 3- and 4-hydroxyl groups of bound mannose form hydrogen bonds with four amino acid side chains that are coordination ligands for this Ca^{2+}. The hydrogen bonds to each of the sugar hydroxyl groups are polar and co-operative: a negatively charged carboxyl group accepts a hydrogen bond from the sugar hydroxyl group, which in turn accepts a hydrogen bond from an asparagine residue. The remaining unshared pair of electrons on each hydroxyl group acts as a coordination ligand for the Ca^{2+}. Thus, the key interactions in the binding site are sensitive to the orientation of the 3- and 4-hydroxyl groups, allowing the selective binding of mannose and other sugars, such as GlcNAc, which have the same disposition of 3- and 4-hydroxyl groups. MBP interacts equally well with mannose and GlcNAc, reflecting the fact that there are no interactions with the 2-substituents that differ between the two sugars. This simple distinction between surfaces rich in mannose-type ligands and those rich in galactose-type ligands is sufficient to direct the innate immune response.

MBP interacts only with the terminal sugar residue in an oligosaccharide chain (Figure 9.3). The dissociation constant for interaction with a high mannose oligosaccharide from a mammalian protein is roughly 1 mM, the same as the dissociation constant for interaction with monosaccharide. High-affinity binding of MBP requires interaction of multiple CRDs with multiple terminal mannose residues. The trimeric structure of MBP presents a cluster of CRDs for interaction with appropriately spaced terminal mannose residues (Figure 9.4). Because the distance between the sugar-binding sites in the trimer is roughly 50 Å, multiple terminal

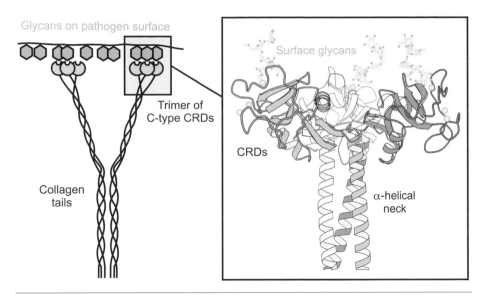

Glycans on pathogen surface

Trimer of
C-type CRDs

Collagen
tails

Surface glycans

CRDs

α-helical
neck

Figure 9.4 Overall structure of MBP, with a close-up view of the trimeric cluster of CRDs. The positions of bound oligosaccharides are shown, indicating how ligands might be projected from a cell surface to interact with the multiple binding sites. [Adapted from: Weis, W.I. and Drickamer, K. (1996). Structural basis of lectin-carbohydrate recognition, *Annual Review of Biochemistry* **65**, 441–473. Based on entries 2MSB and 1RTM in the Protein Databank.]

mannose residues on a single oligosaccharide cannot interact with these multiple binding sites. However, arrays of terminal sugar residues found on the surface of micro-organisms can interact with multiple sites simultaneously. In the ideal case, in which each interaction is independent, the free energy of multiple interactions is additive. As $\Delta G = RT \ln K_D$, the affinities effectively multiply with each additional interaction. A three-way interaction involving three binding sites with dissociation constants of 1 mM can result in an overall dissociation constant of up to 1 nM for a multivalent ligand with appropriately spaced terminal sugar residues.

In order to achieve high-affinity binding to multivalent ligands, the condition of independent binding sites must be met. This means that there must be no distortion of the protein or the sugar ligand. The construction of the MBP trimer imposes important constraints on how this can be achieved. A key feature of the structure is that the spacing between the binding sites is fixed. The trimer is stabilized by a coiled coil of α helices in the neck region and the top portion of this coiled coil interacts extensively with the CRDs to maintain the fixed geometry. This situation contrasts with the presence of a flexible hinge region in antibody molecules, for example. Maintaining the 50 Å distance between binding sites ensures that endogenous high mannose structures cannot bind to MBP with high affinity. In addition to fixing the distance between the sugar-binding sites, the geometry of the MBP trimer establishes the orientation of the binding sites, which all face in the same direction, pointing away from the neck region. This arrangement is obviously suited to the role of MBP in interacting with surfaces of micro-organisms.

9.4   The mannose receptor helps macrophages to internalize pathogens

Like serum MBP, the mannose receptor found on endothelial cells and Kupffer cells in the liver, and on macrophages in other tissues, binds sugars such as mannose and GlcNAc that are frequently found on the surfaces of micro-organisms. There are eight C-type lectin-like domains in the extracellular domain of the mannose receptor, several of which are involved in Ca^{2+}-dependent binding of ligands bearing terminal mannose residues. Carbohydrate binding by the C-type CRDs is pH sensitive, so that ligands are released from the receptor in the acidic environment of the endosomes and directed to the lysosomes for degradation. The mannose receptor is unusual among C-type lectins in having multiple different C-type CRDs in a single polypeptide chain. Binding of the mannose receptor to components of the cell walls of pathogens has been demonstrated and phagocytosis has been shown using cells transfected with the mannose receptor (Figure 9.5). Thus, the mannose receptor plays a part in the innate immune response by helping macrophages bind and internalize pathogens, although it has an additional role in clearance of soluble glycoconjugates.

Several macrophage proteins, including the complement receptor CR3 and the lipopolysaccharide receptor CD14, act in parallel with the mannose receptor in the phagocytosis of many micro-organisms. However, the mannose receptor is the main macrophage protein involved in the first line of defence against *Pneumocystis carinii*. This airborne pathogen is a common cause of pneumonia. A major surface glycoprotein of *Pneumocystis* bears many high-mannose N-linked glycans that serve as ligands for the mannose receptor. Humans are commonly exposed to *Pneumocystis*, but illness is usually only observed when the immune system is compromised. For example, pneumonia caused by *Pneumocystis* can be life threatening in AIDS patients. In these patients, human immunodeficiency virus (HIV) infection causes downregulation of mannose receptor expression, thus interfering with the innate defence against *Pneumocystis*. Other pathogens that bind to the mannose receptor include the pathogenic yeast *Candida albicans*, which infects immunocompromised patients, and the protozoan parasite *Leishmania donovani*, which is responsible for visceral leishmaniasis.

Phagocytosis of most pathogens by macrophages results in killing of the micro-organism following fusion of the phagosomes with lysosomes. However, the mycobacteria that cause tuberculosis and leprosy are able to survive and multiply within macrophages. These organisms make use of the mannose receptor to gain entry into the macrophage, but they prevent fusion of the phagosomes with lysosomes so that they are not destroyed. The mannose receptor binds with high affinity to mannose disaccharides in the lipoarabinomannan that is a major constituent of the capsule of these

⊙ See sections 10.8 and 10.9 for more on the mannose receptor.

Glycobiology and disease
Bacterial infections

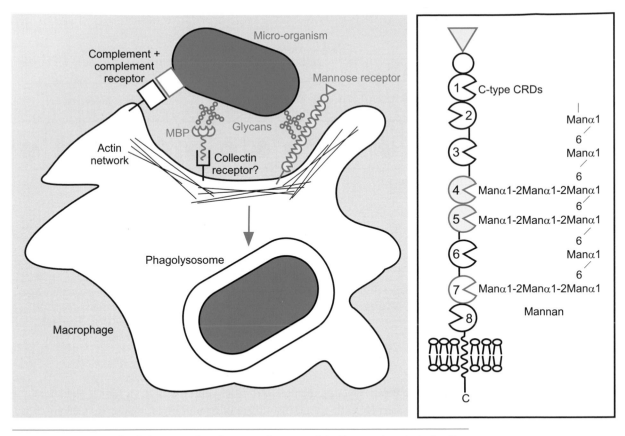

Figure 9.5 Phagocytosis of micro-organisms by macrophages mediated by receptors, including the mannose receptor. Several identification systems probably work in parallel to mediate the interaction between the macrophage and potential pathogens. The repeated, widely spaced terminal sugars on pathogen surfaces allow secondary interactions with C-type lectin-like domains of the mannose receptor beyond the core CRDs, 4 and 5.

mycobacteria. Genetic studies show that a major susceptibility locus for leprosy maps to the gene for the mannose receptor on human chromosome 10. However, the molecular mechanism underlying the increased susceptibility has not yet been determined.

9.5 The selectins are cell adhesion molecules for white blood cells

Circulating leucocytes must interact with endothelial cells lining blood vessels in order to reach the underlying tissues. T- and B-cell homing to peripheral lymph nodes and neutrophil migration to sites of inflammation both involve such interactions. The first step of the leucocyte–endothelium interaction is carried out by the selectin cell adhesion molecules, which are some of the best understood examples of C-type

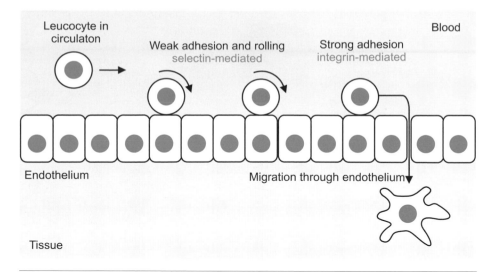

Figure 9.6 Steps in leucocyte interactions with endothelial cells. The selectins mediate the initial, weak binding that leads to rolling of the leucocytes on the endothelial surface. Subsequent integrin-mediated interactions lead to extravasation into the tissue. [Adapted from: Alberts, B., Johnson A., Lewis, J., Raff, M., Roberts, K., and Walter, P. (2002). *Molecular Biology of the Cell* (4th edn). New York: Garland Science.]

lectins. The three selectins mediate an initial rolling phase, in which the leucocyte interacts transiently with glycan ligands, making and breaking contacts (Figure 9.6). These transient binding events facilitate formation of stable protein–protein interactions between the integrins and their counter-receptors. Finally, the leucocytes migrate through the endothelium by passage between endothelial cells. Migration is facilitated by protein–protein interactions that are less well understood.

Homing of lymphocytes to peripheral lymph nodes is a continuous process, as lymphocytes are released, circulate, and then re-enter the nodes. L-selectin on the surface of lymphocytes binds carbohydrate-containing counter-receptors on specialized endothelial cells in the lymph nodes. Although both P-selectin and E-selectin are involved in the migration of neutrophils into sites of inflammation, they operate on different timescales. P-selectin in endothelial cells is stored in intracellular vesicles called Weibel–Palade bodies. When tissue damage occurs, these vesicles rapidly deliver P-selectin to the surface of the endothelial cells in order to attract circulating neutrophils. Once captured, the neutrophils move into the tissue and initiate a protective response. Transcription of E-selectin then increases to allow continued recruitment of neutrophils and macrophage precursors. L-selectin also causes attachment of lymphocytes and additional neutrophils to neutrophils already immobilized on the endothelium, thus providing a further mechanism for leucocyte homing to the inflammatory site.

The phenotypes of knockout mice lacking one or more of the selectins reflect the roles of these proteins in leucocyte trafficking. In the absence of P-selectin, leucocyte rolling on endothelium is significantly reduced and the numbers of leucocytes in the circulation are increased because of their failure to migrate through the endothelium. These phenotypes are substantially enhanced in mice lacking both P- and E-selectin. In addition, the double null mice have increased rates of spontaneous infection, indicating the importance of leucocyte trafficking in the defence against pathogens. Mice lacking L-selectin show impaired lymphocyte homing, but they also show reduced rolling of other leucocytes on inflamed endothelium. This phenotype reflects roles of L-selectin in the recruitment of leucocytes other than neutrophils at these sites. The absence of L-selectin results in a diminished ability to mount an immune response, again confirming the importance of its recruitment and homing functions.

➲ See section 13.7 for more on selectins and cancer.

The three selectins share a common domain organization, with the ligand-binding activity localized to an N-terminal C-type carbohydrate-recognition domain (CRD) (Figure 9.7). This domain is adjacent to an epidermal growth factor-like domain and

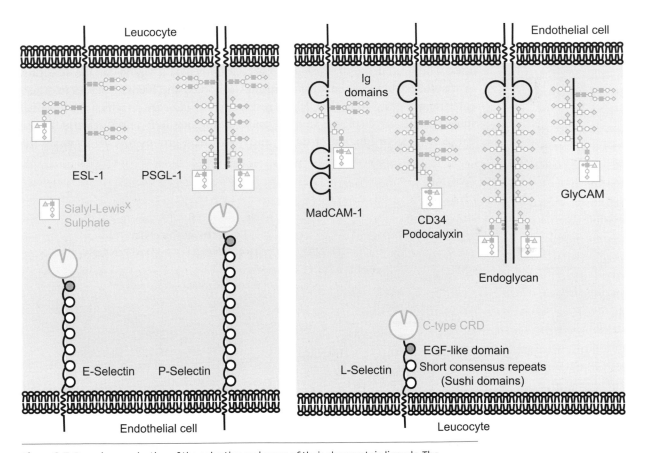

Figure 9.7 Domain organization of the selectins and some of their glycoprotein ligands. The symbols provide an indication of which proteins bear N- and O-linked glycans, but the precise glycan structures and number of attachment sites are not depicted. EGF, epidermal growth factor; ESL, E-selectin ligand; GlyCAM, glycosylated cell adhesion molecule; Ig, immunoglobulin; MadCAM, mucosal addressin cell adhesion molecule; PSGL, P-selectin glycoprotein ligand.

a series of short consensus repeat or Sushi modules, which project the CRD away from the cell surface. The rolling process requires a precise balance between the making and breaking of contacts between the leucocyte and endothelium. This balance is achieved through a variety of factors, some of them intrinsic to the selectin molecules. For example, the rate constants for ligand binding (k_{on}) and release (k_{off}) are rapid. The extended organization of the selectin molecules also allows them to act as mechanical lever arms. In addition to these intrinsic properties, the density and clustering of selectins and their glycan ligands on microvilli that project from the cell surface and interactions of the selectins with the cytoskeleton influence the mechanics of rolling.

9.6 Specific carbohydrate ligands for the selectins interact through extended binding sites on the C-type CRDs

Ligands for the three selectins include the sialyl-Lewis^x tetrasaccharide (Figure 9.8), which is characterized by the presence of terminal fucose and sialic acid residues. The related sialyl-Lewis^a oligosaccharide is also a ligand for the selectins, as are other structures containing terminal sialic acid and fucose residues in different arrangements. Thus, the GlcNAc residue serves as a core on which the terminal sugar residues are displayed. In the ligand-binding sites in E- and P-selectin, fucose is bound by coordination of the 3- and 4-hydroxyl groups to Ca^{2+} in much the

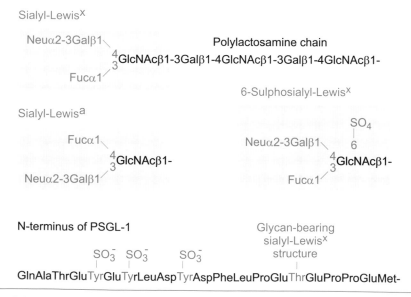

Figure 9.8 Structures of oligosaccharide and glycopeptide ligands for the selectins. Key terminal structures are highlighted in blue.

same way as in the mannose-binding CRDs (Figure 9.9). This monosaccharide-binding site is extended by a series of hydrogen-bonding interactions with the galactose and sialic acid residues. The complete sialyl-Lewis[x] binding site is thus effectively composed of two binding subsites: for fucose and for the galactose–sialic acid disaccharide.

The Lewis[x] portion of this structure has been extensively investigated and is invariably found in a single conformation regardless of whether it is conjugated with sialic acid. The stability of the Lewis[x] conformation results, at least in part, from the favourable van der Waals packing of galactose and fucose. As expected, this conformation remains the same when the ligands are bound to E- or P-selectin. An energy profile viewed in a torsion angle plot reveals that the energies associated with several ψ angles for the sialic acid linkage to galactose are similar. In solution, the lowest energy form is predominant, but alternative conformers are observed in the selectin complexes. Only a small energy cost is paid to move the ligand from its absolute lowest energy conformation. The binding of lowest or nearly lowest energy conformations of oligosaccharides is a general feature of lectin binding to complex ligands, reflecting the fact that substantial distortions of the glycan structure would overwhelm the modest energy gains associated with the limited number of contacts between the protein and sugar.

Although all of the selectins bind sialyl-Lewis[x] and related ligands, each interacts preferentially with these glycans in the context of specific protein carriers (Figure 9.7). The interaction between P-selectin and P-selectin glycoprotein ligand 1 (PSGL-1) from leucocytes has been particularly well characterized. PSGL-1 is

⊙ See section 7.5 for more on the conformation of Lewis[x].

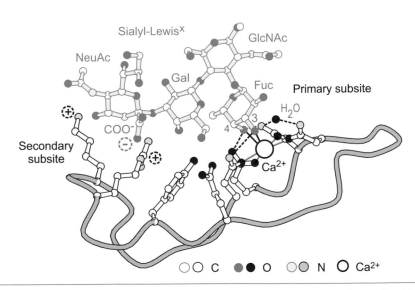

Figure 9.9 Extended binding site in the E-selectin CRD, visualized as primary and secondary binding subsites. GlcNAc serves as a bridge between fucose and galactose. Fucose interacts with the primary binding site, while galactose and sialic acid interact with the positively charged secondary subsite. [Based on entry 1G1T in the Protein Databank.]

⊙ See sections 3.1 and 3.2 for more on mucin-like domains.

a transmembrane protein with an extended mucin-like domain. Most of the O-linked glycans are short, sialylated structures that keep the polypeptide in an elongated conformation. An extended glycan attached to one specific threonine residue near the N-terminus bears the sialyl-Lewis^x structure that is bound by P-selectin (Figure 9.8). However, high-affinity binding requires the presence of sulphated tyrosine residues that lie to the N-terminal side of the key glycosylation site. The sulphate groups interact with a region of positive potential adjacent to the sugar-binding site on the surface of P-selectin. Antibodies to PSGL-1 can block rolling mediated by P-selectin, suggesting that PSGL-1 is the most important P-selectin ligand on leucocytes.

PSGL-1 also serves as a ligand for E-selectin, but the sulphated tyrosine residues are not required for this interaction. E-selectin binds to additional leucocyte surface proteins, including E-selectin ligand 1 (ESL-1). Unlike the other selectin glycoprotein ligands, ESL-1 contains N-linked but not O-linked glycans. The presence of N-acetyllactosamine structures on N-linked glycans allows construction of sialyl-Lewis^x groups like those on the O-linked glycans (Figure 9.8).

Glycoprotein ligands for L-selectin are also mucin-like in character, with multiple sites of O-linked glycan attachment. In mice, secreted glycosylated cell adhesion molecule 1 (GlyCAM-1) functions as a receptor when associated with the plasma membranes of high endothelial venules, but competes for binding when present in serum. In humans and mice, the immunoglobulin superfamily members CD34, endoglycan, and podocalyxin are cell surface receptors for L-selectin. N- and O-linked glycans are attached to an extended mucin-like domain in the extracellular portion of these polypeptides. An additional candidate in mucosal lymph nodes is mucosal addressin cell adhesion molecule 1 (MadCAM-1), which also serves as a receptor for one of the integrins. Glycans that bind L-selectin with high affinity contain sialyl-Lewis^x groups that are further modified by sulphation on the 6-position of GlcNAc residues (Figure 9.8). About 20% of the glycans on GlyCAM-1 and L-selectin-binding forms of CD34 contain these 6-sulpho-sialyl-Lewis^x epitopes. In mice, two GlcNAc-6-O-sulphotransferases, GlcNAc6ST-1 and GlcNAc6ST-2, are involved in generating 6-sulpho-sialyl-Lewis^x. Lymphocyte homing to peripheral lymph nodes is greatly reduced in knockout mice lacking both GlcNAc6ST-1 and GlcNAc6ST-2, indicating the importance of 6-sulpho-sialyl-Lewis^x as a ligand for L-selectin. Endoglycan, but not the other L-selectin glycoprotein ligands, has an N-terminal region that contains sulphated tyrosine residues and a glycan modified with sialyl-Lewis^X, so the interaction of L-selectin with endoglycan is probably very similar to the P-selectin/PSGL-1 interaction. Thus, protein-specific as well as glycan-specific sulphotransferases are required to generate L-selectin ligands.

Glycobiology and disease
Leucocyte adhesion deficiency

Evidence that recognition of fucose-containing ligands by the selectins is important in cellular trafficking in the immune system comes from patients with an extremely rare clinical syndrome, leucocyte adhesion deficiency type II (LAD-II). These patients have high levels of circulating neutrophils and show susceptibility to infection. They are unable to synthesize fucose-containing glycans, including glycan ligands for the selectins, due to a defect in the GDP-fucose transporter that mediates entry of the donor sugar for fucosyltransferases into the lumen of the Golgi apparatus.

➲ See **Box 5.1** for more on LAD-II.

9.7 C-type lectins participate in the process of antigen presentation

C-type lectins play two different roles in the presentation of antigens to T cells, which is the first step in the development of an antibody-mediated or cytotoxic T-cell-mediated adaptive immune response (Figure 9.10). Several of the receptors that scavenge for potential foreign substances are members of the C-type lectin family. Using these receptors, dendritic cells that are resident in peripheral tissues (such as skin and the lining of the intestine) capture protein and glycoprotein antigens. The

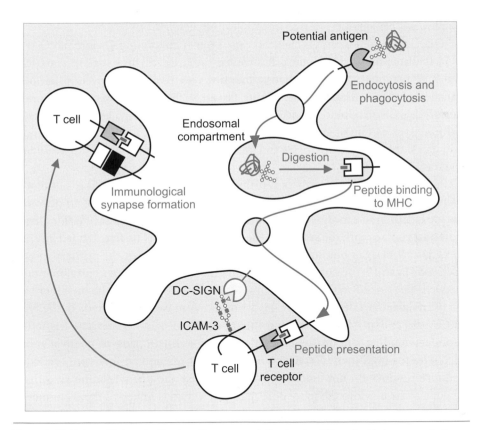

Figure 9.10 Roles of C-type lectins in antigen presentation by dendritic cells. Various receptors, including endocytic C-type lectins such as the mannose receptor, mediate internalization of potential antigens, which are degraded and presented on the dendritic cell surface by molecules of the MHC. Probing of the dendritic cell surface by T cells is facilitated by the interaction between DC-SIGN and ICAM-3. Interaction of a T-cell receptor with one of the presented antigens leads to formation of an immunological synapse and stimulation of the T cell.

internalized antigens are digested to produce peptide fragments and, after the dendritic cells migrate to lymph nodes, fragments of the antigens return to the cell surface in complex with molecules of the major histocompatibility complex (MHC). Specific peptide–MHC complexes can interact with T-cell receptors, but only a small fraction of the total T-cell population express receptors appropriate for a particular peptide. Therefore, T cells must sample the surfaces of dendritic cells to determine if cognate peptide–MHC complexes are present. The sampling process is facilitated by receptors that bring the surfaces of dendritic cells and T cells near to each other. Carbohydrates on intercellular adhesion molecule 3 (ICAM-3) on lymphocytes are bound by the C-type lectin DC-SIGN (dendritic cell-specific ICAM-3-grabbing non-integrin). If there is an appropriate match between the T-cell receptor and a peptide–MHC complex on the dendritic cell surface, this initial interaction can lead to the formation of an immunological synapse, in which further adhesion molecules consolidate the adhesion event and initiate intracellular signalling events leading to lymphocyte activation. If the T cell does not encounter an appropriate peptide–MHC complex, it is presumably released to sample other dendritic cells.

DC-SIGN is a tetrameric transmembrane protein containing C-terminal C-type CRDs that are projected from the cell surface by an α-helical neck region. The CRDs bind high mannose, N-linked oligosaccharides, and a variety of glycans that bear branched terminal structures containing fucose such as the Lewis^x trisaccharide. Thus, DC-SIGN may bind to specific fucosylated glycans on ICAM-3. However, it also functions as one of the scavenging receptors, binding to clusters of mannose residues on the surfaces of viruses and fucose-containing glycans that are abundant on parasitic nematodes.

The CRD of DC-SIGN utilizes a primary monosaccharide binding site that is essentially identical to that in MBP. However, while the remainder of an oligosaccharide ligand points away from the surface of MBP, bound oligosaccharides make additional contact with the surface of the CRD of DC-SIGN in an extended binding site (Figure 9.11). For binding high-mannose oligosaccharides, the primary monosaccharide ligand is the outer branched trimannose structure. The central mannose residue is positioned so that the two branching mannose residues lie on either side of a phenylalanine side chain. The α1-3-linked mannose residue binds in the typical Ca^{2+}-dependent binding site and the α1-6-linked mannose residue fits into a secondary binding site on the other side of the aromatic ring of phenylalanine. Selectivity for high mannose structures is a result of the fact that core branched mannose residues do not bind to the CRD because the phenylalanine side chain creates a steric block to the inner core mannose which is linked as the β anomer. It is interesting that the mannose residue in the primary binding site of DC-SIGN does not need to be terminal, because it can be extended with an α1-2-linked mannose residue commonly found in larger high mannose structures. In the case of the fucose-containing ligands such as Lewis^x trisaccharide, fucose is accommodated in the primary binding site, but additional contacts are made by the adjacent galactose residue.

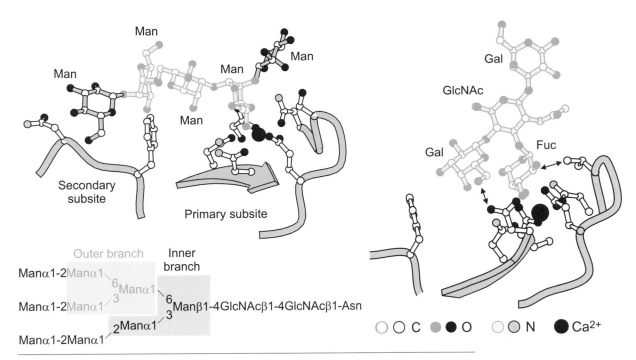

Man
Man
Man
Man
Man
Man
Man

Secondary
subsite

Primary subsite

Gal
GlcNAc
Gal
Fuc

Outer branch   Inner
branch
Manα1-2Manα1
6 Manα1
3
Manα1-2Manα1
6 Manβ1-4GlcNAcβ1-4GlcNAcβ1-Asn
3
2 Manα1
Manα1-2Manα1

○ ○ C   ● ● O   ○ ○ N   ● Ca2+

Figure 9.11 Extended binding sites in DC-SIGN showing how a branched trimannose structure in the outer arm of a high-mannose oligosaccharide is arranged in two subsites on either side of a phenylalanine side chain and how branched oligosaccharides with terminal fucose residues make secondary contacts with the surface of the CRD. [Based on entry 1K9J in the Protein Databank.]

9.8 DC-SIGN enhances HIV infection of T cells

HIV utilizes DC-SIGN to facilitate infection of T cells, which initiates AIDS. The receptor binds to the viral surface glycoprotein gp120, which is rich in N-linked high mannose glycans. Thus, dendritic cells that encounter HIV in the vaginal or intestinal mucosa can carry it to lymph nodes, where it migrates to the surface of T cells (Figure 9.12). Binding to, and infection of, the T cells is mediated by CD4 and co-receptors. The pool of virus on the surface of dendritic cells can remain latent for extended periods, contributing to the difficulty in eliminating the virus. Although DC-SIGN can mediate endocytosis and degradation of soluble glycoproteins, the virus is either not internalized or is returned to the surface intact. A related receptor, DC-SIGNR (DC-SIGN-related protein), found on endothelial cells of liver, lymph node, and placenta also binds HIV and several other viruses (including hepatitis C virus and Ebola virus) through high-mannose oligosaccharides. The presence of DC-SIGNR on maternal and foetal cells in the placenta suggests that it may facilitate vertical transmission of HIV.

Although the roles of DC-SIGN are much less well understood than the roles of the selectins, it is interesting to compare some of their features. Both the selectins and DC-SIGN participate in reversible adhesion events between two cells, or

Glycobiology and disease
Human immunodeficiency virus

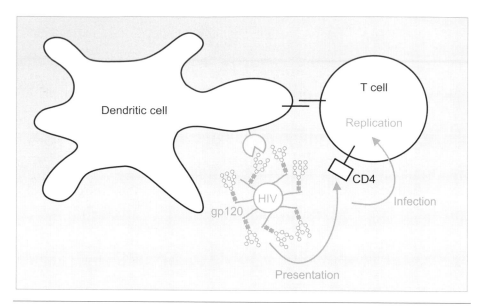

Figure 9.12 DC-SIGN and DC-SIGNR can enhance infection of T cells by HIV. The virus binds to the dendritic cell surface, but the interaction does not lead to infection. Virus on the surface of the dendritic cell transfers efficiently to T cells that have appropriate cell surface molecules such as CD4 and chemokine receptors that can mediate infection. A single glycan on each gp120 molecule is used to represent multiple N-glycosylation sites bearing various high-mannose oligosaccharides.

between a cell and a virus. However, these interactions take place in rather different contexts, in the presence of shear flow or under static conditions, and the selectins bind to complex glycan ligands that are expressed on a limited number of glycoproteins, while DC-SIGN binds to relatively common types of glycans.

9.9 I-type lectins are composed of immunoglobulin-like domains

Cell surface receptors that contain immunoglobulin-like protein domains participate in many different adhesion and signalling events by binding to protein or carbohydrate ligands. Receptors that bind carbohydrate ligands through immunoglobulin superfamily domains are referred to as I-type lectins. Comparison of the sequences of I-type CRDs with other immunoglobulin superfamily domains shows that one set of sialic acid-binding domains form a particularly closely related group. Members of this group are called siglecs (sialic acid-binding, immunoglobulin-like lectins) and are the best characterized of the I-type lectins.

The 14 human siglecs are transmembrane proteins with broadly similar overall domain organization (Figure 9.13). Immunoglobulin-like domains related to immunoglobulin variable (V-set) and constant (C-set) domains serve two functions in these proteins. The C-set domains are spacers that project the N-terminal V-set domain away from the cell surface. The V-set domain contains the sialic acid-binding site. In

Box 9.1   157

BOX 9.1   Glycotherapeutics: *Drugs and antibodies that prevent HIV infection mimic recognition by DC-SIGN*

The interaction of DC-SIGN with gp120 on the surface of HIV involves two unusual properties of the viral glycoprotein compared with most human cell surface glycoproteins: the close spacing of multiple high-mannose oligosaccharides on the protein and the relatively large size of the glycans, which are mostly Man_7, Man_8, and Man_9 forms. Glycosylation provides a mechanism to shield conserved portions of the gp120 polypeptide from the immune system and also allows the virus to use DC-SIGN as a means of trafficking to lymph nodes. However, both the unusual nature of the glycans and their high density are aspects of the virus that may be exploited therapeutically.

Cyanovirin-N, a small protein that binds with high affinity to Man_8 and Man_9-type glycans, was discovered in a broad screen of potential antimicrobial compounds. In tissue culture, it blocks HIV infection of T cells by preventing gp120 from binding to the CD4 receptor. Cyanovirin-N binds with high affinity to the Man_8 and Man_9 glycan structures found on gp120. Although it competes with DC-SIGN for binding to Man_8 and Man_9 glycans, the parts of such glycans that are recognized appear to be different: the terminal Manα1-2Man structures form the critical binding determinants for cyanovirin-N, while the outer branching structure is most important for DC-SIGN binding.

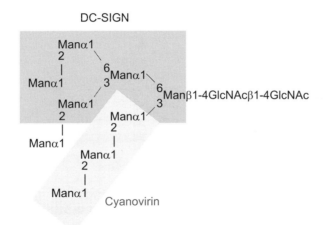

Preclinical testing in tissue culture models of vaginal and rectal infection confirms that cyanovirin is able to block HIV infection and recent studies in macaques show that a vaginal gel formulation can prevent infection. The fact that cyanovirin can be used topically is a way around the problems usually associated with delivery of protein therapeutics. In this situation, cyanovirin probably works both by blocking infection of T cells locally and also by preventing DC-SIGN on dendritic cells from taking up virus and presenting it to T cells in lymph nodes.

Essay topics

- Discuss the importance of multivalent binding of oligosaccharides in the interactions of gp120 of HIV with cyanovirin.
- Describe why cyanovirin, although it is a protein, is a strong candidate for therapeutic applications.

Lead references

• Botos, I. and Wlodawer, A. (2003). Cyanovirin-N: a sugar-binding antiviral protein with a new twist, *Cellular and Molecular Life Sciences* **60**, 277–287.

• Barrientos, L.G., Lasala, F., Delgado, R., Sanchez, A., and Gronenborn, A.M. (2004). Flipping the switch from monomeric to dimeric CV-N has little effect on antiviral activity, *Structure* **12**, 1799–1807.

• Tsai, C.-C., Emau, P., Jiang, Y., Agy, M.B., Shattock, R.J., Schmidt, A., Morton, W.R., Gustafson, K.R., and Boyd, M.R. (2004). Cyanovirin-N inhibits AIDS virus infection in vaginal transmission models, *AIDS Research and. Human Retroviruses* **20**, 11–18.

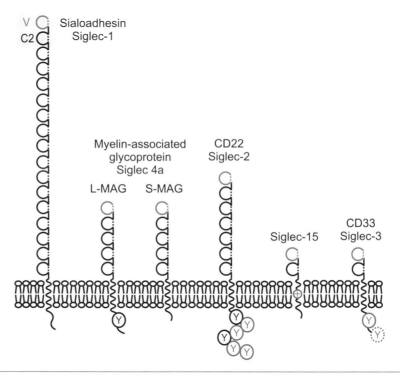

Figure 9.13 Organization of some of the siglecs, with ligand-binding V-set immunoglobulin domains indicated in blue and spacer domains of the C2 subset in black. Single genes encoding sialoadhesin, CD22, MAG, and siglec-15 are found in both mouse and humans. CD33 is the prototype for a subfamily of closely related proteins for which there are different sets of genes in mouse and humans. Nine CD33-related siglecs have been identified in humans and five in mice. Tyrosine-containing sequence motifs in the cytoplasmic tails of some of the siglecs are indicated by circles marked with 'Y'. Those that fit the consensus for immunotyrosine-based inhibition motifs (ITIMs) are highlighted in blue, with a dotted circle representing a variant motif common to the CD33-related siglecs. Siglec-15 does not contain ITIMs, but can regulate signalling by association via a lysine residue in its membrane-spanning domain with the ITIM-containing adaptor protein DAP-12. [Adapted from: Crocker, P.R. and Varki, A. (2001). Siglecs, sialic acid and innate immunity, *Trends in Immunology* **22**, 337–342.] MAG, myelin-associated glycoprotein.

each siglec, the N-terminal V-set domain is linked to the adjacent C-set domain by an unusual inter-module disulphide bond. The presence of this bond reflects the close evolutionary relationship between the siglecs, in spite of differences in the number of spacer domains. Most of the siglecs are present on cells of the immune system and our knowledge of these siglecs can be illustrated by discussion of sialoadhesin and CD22. Myelin-associated glycoprotein is expressed in the nervous system.

➲ See section 12.6 for more on MAG.

9.10 Siglecs are adhesion and signalling receptors on cells in the immune system

The extended structure of sialoadhesin, the largest member of the siglec family, ensures that the terminal, sialic acid-binding domain is located well above most of the proteins and sugars at the cell surface, so it is accessible for interaction with other cells or components of the extracellular matrix. Sialoadhesin binds preferentially to terminal *N*-acetylneuraminic acid (NeuAc) residues in α2-3 linkage to galactose, but shows little selectivity for parts of glycans beyond this terminal structure and in fact will also bind to NeuAcα2-3GalNAc and NeuAcα2-6Gal with low affinity (Figure 9.14). This relatively broad ligand-binding specificity is explained by the structure of the NeuAc-binding site in the terminal immunoglobulin-like domain,

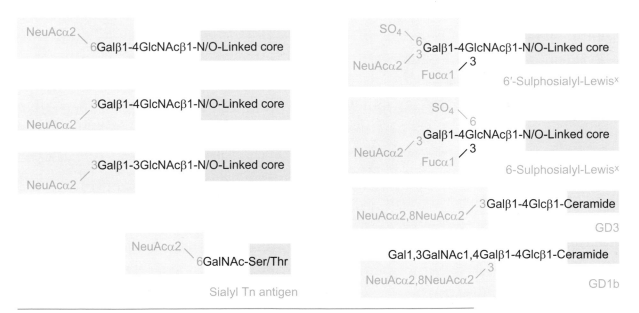

Figure 9.14 Sialic acid-containing ligands for some of the siglecs. Sialoadhesin and CD22 probably bind common N- and O-linked glycans attached to glycoproteins. Some of the CD33-related siglecs bind to specific ligands such as 6'-sulphosialyl Lewis[x] (human siglec 8), 6-sulphosialyl Lewis[x] (human siglec 9), and disialyl motifs found in GD3 and GD1b (human siglec 7).

in which the binding pocket accommodates NeuAc with great selectivity but does not form extensive interactions with the remainder of its sialylated ligands. In sialoadhesin, there are two direct hydrogen bonds between the carboxyl group of the sialic acid ligand and the side chain of Arg 97 in the binding site (Figure 9.15). The interactions in the binding site are arranged to read out the positions of each substituent around the sialic acid ring. In addition to the arginine–carboxyl interactions, there are hydrogen bonds with hydroxyl groups at positions 4, 7, 8, and 9, and with the nitrogen attached to C5. Tryptophan side chains make hydrophobic interactions with the hydrocarbon backbone of the C7–C9 glycerol side chain, and with the methyl group of the 5-*N*-acetyl substituent. These interactions make the binding site highly selective for sialic acid, although the affinity for the monosaccharide is comparable with the affinity of C-type CRDs for monosaccharide ligands, in the millimolar range. Some of the CD33-related siglecs show greater selectivity for more complex sialic acid-containing ligands (Figure 9.14).

The precise functions of sialoadhesin are not known. Given its affinity for relatively common terminal sugar structures, it is interesting that sialoadhesin shows selectivity in binding to specific cell types such as granulocytes. Selectivity may arise because some cells express glycoproteins that present high concentrations of the appropriate ligands. Such ligands include mucins and proteins containing mucin-like domains that contain multiple sialylated, O-linked glycans. For example, sialoadhesin

○ See sections 3.1 and 3.2 for more on mucin-like domains.

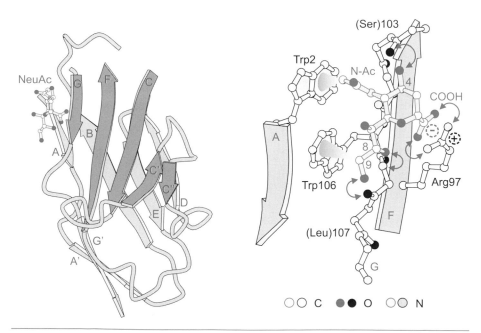

Figure 9.15 Overall fold and monosaccharide-binding site of the CRD from sialoadhesin. The binding site is located at the edge of the two β-sheets that make up the immunoglobulin domain and is arranged to interact with each of the substituents on the ring of the sialic acid residue. [Based on entry 1QFO in the Protein Databank.]

binds to the O-glycosylated region of the cell surface molecule CD43 as well as to P-selectin glycoprotein ligand 1. These glycoproteins, found on the surfaces of T lymphocytes, thus facilitate macrophage/T-cell interactions. The pattern of sialoadhesin expression suggests that it may play a part in multiple macrophage adhesion processes, because it is found on macrophages that are resident in tissues as well as on macrophages that infiltrate tumours and sites of inflammation.

The cytoplasmic domain of CD22, a siglec found on B cells, contains multiple tyrosine-based sequence motifs that can be phosphorylated and either activate or inhibit intracellular signalling cascades. Even though CD22 is selective for α2-6-linked sialic acid, this ligand is widely distributed on soluble and membrane-bound glycoproteins. Thus, it is difficult to picture a simple ligand–receptor relationship such as that between insulin and the insulin receptor. It seems more likely that signalling results from the engagement of multiple CD22 molecules by an array of sialic acid residues on an adjacent cell surface. In spite of the presence of both activating and inhibiting motifs in CD22, the phenotype of knockout mice suggests that the inhibitory function is predominant, because mice that lack CD22 show increased sensitivity to bacterial lipopolysaccharide and an increased tendency to produce auto-antibodies. Sialic acid is common on mammalian cells but rare on the surface of bacterial and fungal pathogens, so a possible function for CD22 is inhibition of B cell activation when confronted with sialic acid-bearing cells as a means to prevent self-reactivity. In this case, recruitment of SHP phosphatases could lead to inhibition of signalling from the B cell antigen receptor (Figure 9.16). Members of the CD33-related subgroup of siglecs, which

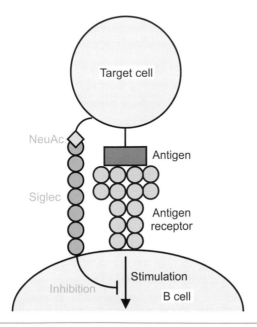

Figure 9.16 Possible function of CD22 in inhibiting reactivity with sialic acid-bearing cells. Binding of CD22 to sialic acid-rich surfaces may lead to inhibition of signalling from the antigen receptor.

are differentially expressed on white blood cells, also contain inhibitory motifs that can recruit SHP phosphatases, suggesting roles in regulation of immune responses.

CD22 also functions as an adhesion receptor and mediates homing of B cells to bone marrow. Because CD22 contains fewer spacer immunoglobulin domains than sialoadhesin, its binding site is not projected so far from the surface and on many B cells it is masked by binding to sialic acid-containing glycoproteins present nearby in the B cell plasma membrane. However, it is postulated that the CD22 binding site on some cells can become unmasked, and that these cells will selectively migrate to the bone marrow. The precise target ligands for CD22 on bone marrow endothelium are not known. CD22 shows strong selectivity for NeuAc residues in α2-6 linkage to galactose or GalNAc, but shows highest binding affinity for a ligand with 6-sulphated GlcNAc, α2-6-sialylated 6-sulpho-*N*-acetyllactosamine (NeuAcα2-6Galβ1-4[6S]GlcNAc). Glycoproteins containing α2-6-sialylated 6-sulpho-*N*-acetyllactosamine are found on endothelial cells of lymphoid tissue and may serve as ligands for CD22 in B cell homing.

9.11   Extracellular galectins have roles in cell adhesion and cell signalling

⦿ See section 10.10 for more on galectins in the nucleus.

The galectins are a family of soluble proteins that bind β-galactosides. They have a wide cell and tissue distribution, and proposed roles in the nucleus. However, galectins also reach the cell surface by a poorly understood process in which they become enclosed in membrane vesicles that fuse with the plasma membrane. Secreted galectins can mediate or modulate cell–cell interactions, cell–matrix adhesion, and transmembrane signalling (Figure 9.17). Expression and secretion of the galectins is regulated, suggesting that they function at specific times during development. The phenotypes of knockout mice lacking individual galectins also indicate that these proteins function in specific developmental processes. For example, mice lacking galectin 1 appear superficially normal, but some of their primary sensory neurons in the olfactory system do not project axons to the correct sites in the olfactory bulb.

Numerous potential ligands for the galectins are found outside cells. They bind tightly to the polylactosamine chains that are commonly found on cell surface proteins and in the extracellular matrix. Different galectins show distinct specificities for oligosaccharides, although galactose is a necessary constituent of all ligands. Cell surface glycoproteins and matrix components bound by the galectins include integrins, fibronectin, laminin, and tenascin. The galectins do not have hydrophobic membrane anchors, so they mediate adhesive events by cross-linking two glycans. The bivalent binding activity required to perform this function is achieved in different ways in different subgroups of galectins (Figure 9.17). Many galectin polypeptides contain a single CRD, but these form dimers that are stabilized by direct interactions between the CRDs. In contrast, members of the tandem galectin subgroup are intrinsically divalent because they contain two CRDs in a single

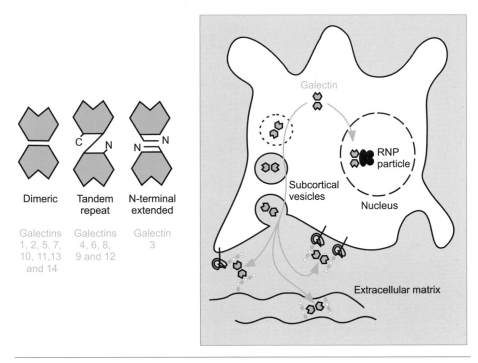

Figure 9.17 Summary of galectin structure and function. Different mechanisms of forming divalent structures are shown on the left. Two proposed roles for galectins, in the nucleus and at the cell surface, are shown on the right. A possible mechanism for the secretion of galectins made in the cytoplasm is shown, involving sequestration within vesicles which then fuse with the plasma membrane. Secreted galectins cross-link glycoconjugates at the cell surface and in the extracellular matrix. RNP, ribonucleoprotein.

polypeptide chain. A third subgroup of galectins, represented by galectin 3 in mammals, can dimerize through an N-terminal extension. Dimerization of galectin 3 is concentration dependent, so it is likely to exist in both monomeric and dimeric forms *in vivo*. Galectins in monomeric form could modulate cell adhesion by blocking binding sites for other adhesive proteins.

Although binding of galectins to galactose and simple derivatives such as α-methyl galactoside can be observed, physiologically significant binding probably requires the presence of a lactose (Galβ1-4Glc) or an *N*-acetyllactosamine (Galβ1-4GlcNAc) disaccharide. The binding site in the galectins is essentially one continuous surface that binds along the side of both sugar residues in the ligand (Figure 9.18). In addition to a hydrophobic packing interaction between the B face of the galactose residue and a conserved tryptophan residue, there are hydrogen bonds to both sugar residues, some involving the characteristic 3- and 4-hydroxyl groups. The greater affinity of galectins for Galβ1-4GlcNAc compared with lactose probably results from van der Waals interactions with the 2-acetamide substituent, although these interactions do not involve hydrophobic packing against an aromatic ring as observed in some of the previous examples. The 1-hydroxyl group of the GlcNAc

LIVERPOOL JOHN MOORES UNIVERSITY
LEARNING SERVICES

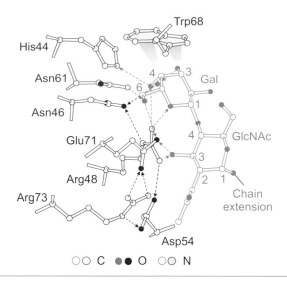

Figure 9.18 Galβ1-4GlcNAc binding site in galectin 1, showing interactions with both sugar residues. Accessibility of the 1-hydroxyl group of GlcNAc and the 3-hydroxyl group of galactose allows elongation of a polylactosamine chain in both directions. [Based on entry 1HLC in the Protein Databank.]

residue points away from the surface of the galectin, so that *N*-acetyllactosamine groups at the non-reducing ends of glycans can be accommodated in the binding site. This arrangement is consistent with the particularly high affinity of the galectins for polylactosamine chains that terminate in this structure.

Interestingly, the geometry of the galectin oligomers is very different to the geometry of mannose-binding protein, as the two binding sites extend from opposite ends of the dimer (Figure 9.19). This alternative geometry allows the galectins to bridge between two glycans. Such cross-linking activity can be associated with links between cell surfaces and the surrounding matrix or with establishing the distance between glycoproteins in the plasma membrane.

9.12 Galectins modulate activation of T cells and control cell survival by triggering or inhibiting apoptosis

In addition to their roles in development, the galectins function in the immune system. For example, galectin 3 cross-linking suppresses activation of T cells by linking T-cell receptors and other glycoproteins into a lattice on the cell surface (Figure 9.20). Mice that lack GlcNAc-transferase V, and therefore are deficient in polylactosamine chains on N-linked glycans, show increased susceptibility to autoimmune diseases and have hypersensitive T cells. The increased sensitivity of the T cells is due to enhanced clustering of T-cell receptors on the carbohydrate-deficient cells. A similar phenotype with enhanced T-cell receptor clustering can be induced in wild-type

⊘ See section 2.6 for more on GlcNAc-transferase V.

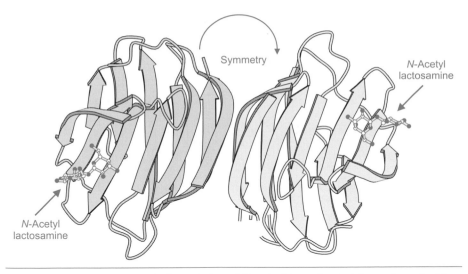

Figure 9.19 Structure of a galectin dimer. The bound disaccharides are located at opposite ends of the molecule, allowing it to act as a bridge between glycans. [Based on entry 1HLC in the Protein Databank.]

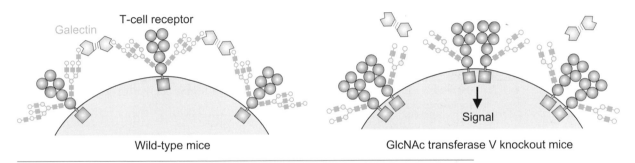

Figure 9.20 Suppression of T-cell receptor activation by galectin 3. Cross-linking of glycans by galectin 3 keeps the T-cell receptor molecules spaced apart. In the GlcNAc-transferase V knockout mouse, the polylactosamine chains that serve as galectin 3 ligands on the T-cell receptor are not synthesized, so the T-cell receptor molecules can cluster and activate the lymphocyte. In the diagram, the polylactosamine chains are abbreviated in length.

cells by pretreatment with lactose, a galectin ligand, but not with other sugars that are not bound by galectins. Thus, the absence of carbohydrate ligands for galectin 3 on T-cell glycoproteins or the presence of competitors of galectin binding prevent lattice formation, allowing T-cell receptors to cluster inappropriately and signal even in the absence of antigen presentation.

Galectins are also involved in regulation of the immune system through their ability to trigger or inhibit programmed cell death (Figure 9.21). Two galectins, galectin 1 and galectin 9, a tandem type galectin, can induce apoptosis of T cells by cross-linking ligands in the plasma membrane. Apoptosis is important at two stages of T cell development. During maturation and selection of T cells in the thymus, thymocytes that express self-reactive receptors and thymocytes that fail to rearrange

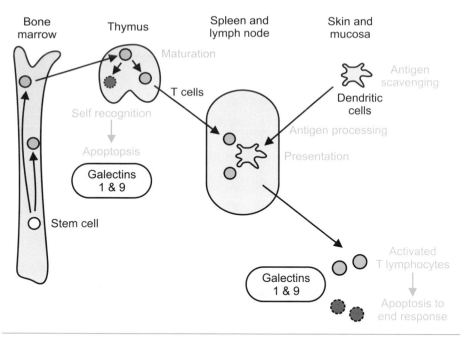

Figure 9.21 Functions of galectins in apoptosis of T cells. Thymocytes migrate from the bone marrow to the thymus, where self-reactive cells are eliminated by apoptosis. In the spleen and lymph nodes, a subset of the T cells recognizes foreign antigens presented by dendritic cells. After these cells have been activated by engaging antigen at sites of inflammation in peripheral tissues, the response is terminated by apoptosis. Apoptosis in the thymus and at sites of inflammation is stimulated by galectins 1 and 9.

a functional T-cell receptor undergo apoptosis. Later, when T cells have migrated to the peripheral tissues and become activated, antigen-specific T cells are killed by apoptosis to bring an end to the response to a particular antigen. Many factors are involved in the process of apoptosis, but it is likely that galectin 1 and galectin 9 contribute to stimulating cell death at both of these stages of T cell development. Both galectins are expressed in the thymus, where T cell selection occurs, as well as in peripheral tissues where activated antigen-specific T cells are found and both galectins have been shown to kill immature thymocytes as well as activated T cells. Galectin 1 binds to several proteins expressed on T cells, but CD7, CD43, and CD45 are the ligands involved in stimulating apoptosis. Galectin 1-mediated cross-linking of core 2 type O-linked glycans on these proteins causes them to cluster in the membrane, initiating intracellular signalling pathways that trigger apoptosis. Expression of core 2 type O-glycans that serve as the ligands for galectin 1 on T cells is developmentally regulated by changes in expression of glycosyltransferases. Core 2 glycans are found on cell-surface glycoproteins of immature thymocytes undergoing selection and on glycoproteins of activated T cells, making them susceptible to galectin 1-induced apoptosis. Mature thymocytes and naive T cells do not express core 2 O-glycans.

⊘ See **section 12.8** for more on glycans and T cell development.

In contrast to galectin 1 and galectin 9, which stimulate apoptosis, galectin 3 can inhibit cell death. Experiments with several types of transfected cell lines, including transfected human leukaemia T cell lines, show that cells that overexpress galectin 3 have higher rates of growth than control cells and are resistant to apoptosis. However, inhibition of apoptosis is caused by galectin 3 acting intracellularly, rather than by cross-linking of glycoproteins at the cell surface. It is interesting that many tumour cells express high levels of galectin 3, suggesting that this galectin might contribute to the uncontrolled growth of these cells.

SUMMARY

Cell adhesion and signalling is mediated by structurally diverse lectins that recognize diverse glycans. One common feature of these systems is that they often involve transient interactions between cells that are moving relative to each other. The processes mediated by these cell surface lectin–glycan interactions are evolutionarily recent compared with the intracellular processes discussed in Chapter 10. Mice lacking individual lectins of the types discussed in this chapter are viable, but suffer from developmental or immunological abnormalities. The phenotypes of these mice demonstrate the importance of lectin recognition events in the precise orchestration of cell–cell interactions in mammals.

KEY REFERENCES

Bleijs, D.A., Geijtenbeek, T.B.H., Figdor, C.G., and van Kooyk, Y. (2001). DC-SIGN and LFA-1: a battle for ligand, *Trends in Immunology* **22**, 457–463. This is a short review of the evidence for a role of DC-SIGN in mediating adhesion of T cells to dendritic cells through binding to ICAM-2 and ICAM-3.

Crocker, P.R., Paulson, J.C., and Varki, A. (2007). Siglecs and their roles in the immune system, *Nature Reviews Immunology* **7**, 255–266. This review, and the paper by McMillan and Crocker (2008) (see below), provide a good overview of the biochemistry and biology of the siglecs.

Epperson, T.K., Patel, K.D., McEver, R.P., and Cummings, R.D. (2000). Noncovalent association of P-selectin glycoprotein ligand-1 and minimal determinants for binding to P-selectin, *Journal of Biological Chemistry* **275**, 7839–7853. Key experiments defining the interaction between P-selectin and its main physiological ligand are presented.

Garner, O.B. and Baum, L.G. (2008). Galectin-glycan lattices regulate cell-surface glycoprotein organization and signalling, *Biochemical Society Transactions* **36**, 1472–1477. A critical review of the evidence for biological functions of galectins due to glycoprotein cross-linking.

Ip, W.K.E., Takahashi, K., Ezekowitz, R.A., and Stuart, L.M. (2009). Mannose-binding lectin and innate immunity, *Immunological Reviews* **230**, 9–21. An overview of the properties of mannose-binding protein and its functions in the immune system.

Khoo, U.-S., Chan, K.Y.K., Chan, V.S.F., and Lin, C.L.S. (2008). DC-SIGN and L-SIGN: the SIGNs for infection, *Journal of Molecular Medicine* **86**, 861–874. The evidence for roles of DC-SIGN and DC-SIGNR in facilitating viral infection is reviewed.

May, A.P., Robinson, R.C., Vinson, M., Crocker, P.R., and Jones, E.Y. (1998). Crystal structure of the N-terminal domain of sialoadhesin in complex with 3'-sialyllactose at 1.85 Å resolution, *Molecular Cell* **1**, 719–728. The molecular basis for sialic acid binding by I-type CRDs of the siglecs is presented.

McMillan, S.J. and Crocker, P.R. (2008). CD33-related sialic-acid-binding immunoglobulin-like lectins in health and disease, *Carbohydrate Research* **343**, 2050–2056. See Crocker *et al.* (2007) above.

Nitschke, L. (2009). CD22 and Siglec-G: B cell inhibitory receptors with distinct functions, *Immunological Reviews* **230**, 128–143. A detailed review of the roles of CD22 in B cell signalling.

Rini, J.M. (1995). Lectin structure, *Annual Review of Biophysics and Biomolecular Structure* **24**, 551–577. This paper, and that by Weis and Drickamer (1996) (see below), provide detailed reviews of animal lectin structures determined by crystallography, and the principles underlying protein–carbohydrate interactions.

Rosen, S.D. (2006). Homing in on L-selectin, *Journal of Immunology* **177**, 3–4. A short commentary highlighting the important experiments that lead to the discovery of l-selectin and the other selectins and defined their roles in leucocyte trafficking.

Somers, W.S., Tang, J., Shaw, G.D., and Camphausen, R.T. (2000). Insights into the molecular basis of leukocyte tethering and rolling revealed by structures of P- and E-selectin bound to SLe^x and PSGL-1, *Cell* **103**, 467–479. Crystal structures of the C-type CRDs of two selectins in complex with sialyl-Lewis^x are described.

Sperandio, M., Gleissner, C.A., and Ley, K. (2009). Glycosylation in immune cell trafficking, *Immunological Reviews* **230**, 97–113. A detailed discussion of the roles of selectins and their ligands, and other potential lectin–glycan interactions in trafficking of immune cells.

Stahl, P.D. and Ezekowitz, R.A.B. (1998). The mannose receptor is a pattern recognition receptor involved in host defense, *Current Opinion in Immunology* **10**, 50–55. This paper provides an overview of the properties of the mannose receptor and its possible functions in the immune system.

Steinman, R.M. (2000). DC-SIGN: a guide to some mysteries of dendritic cells, *Cell* **100**, 491–494. This short commentary provides a good introduction to the functions of dendritic cells in the immune system and the possible roles of DC-SIGN.

Taylor, M.E. and Drickamer, K. (2007). Paradigms for glycan-binding receptors in cell adhesion, *Current Opinion in Cell Biology* **19**, 572–577. A critical review of our current understanding of how different glycan-binding receptors are involved in cell adhesion.

Uchimura, K., Gauger, J.-M., Singer, M.S., Tsay, D., Kannagi, R., Muramatsu, T., von Andrian, U.H., and Rosen, S.D. (2005). A major class of L-selectin ligands is eliminated in mice deficient in two sulfotransferases expressed in high endothelial venules, *Nature Immunology* **6**, 1105–1113. Analysis of the phenotypes of knockout mice for two sulphotransferases shows the importance of sulphated sialyl-Lewisx in lymphocyte homing.

Van der Merwe, P.A. (1999). Leukocyte adhesion: high-speed cells with ABS, *Current Biology* **9**, R419–R422. The kinetic properties of selectin–ligand interactions in leucocyte adhesion are reviewed.

Van Kooyk, Y. and Geijtenbeek, T.B.H. (2003). DC-SIGN: escape mechanism for pathogens, *Nature Reviews Immunology* **3**, 697–709. This review gives details of the interactions of DC-SIGN with pathogenic micro-organisms.

Van Kooyk, Y. and Rabinovich, G.A. (2008). Protein–glycan interactions in the control of innate and adaptive immune responses, *Nature Immunology* **9**, 593–601. A detailed overview of the roles of the multiple different types of glycan-binding proteins in regulation of immune responses.

Wallis, R., Shaw, J.M., Uitdehaag, J., Chen, C.-B., Torgersen, D., and Drickamer, K. (2004). Localization of the serine protease-binding sites in the collagen-like domain of mannose-binding protein: indirect effects of naturally occurring mutations on protease binding and activation, *Journal of Biological Chemistry* **279**, 14065–14073. Key experiments that show the molecular basis for the immunodeficiency caused by mutations in human mannose-binding protein are presented.

Weis, W.I. and Drickamer, K. (1996). Structural basis of lectin–carbohydrate interaction, *Annual Review of Biochemistry* **65**, 441–473. See Rini (1995) above.

Weis, W.I., Taylor, M.E., and Drickamer, K. (1998). The C-type lectin superfamily in the immune system, *Immunological Reviews* **163**, 19–34. The structures and functions of C-type lectins with roles in the immune system are reviewed. Mannose-binding protein, other collectins, the mannose receptor, and the selectins, as well as C-type lectin-like proteins on natural killer cells are included.

Yang, R.-Y., Rabinovich, G.A., and Liu, F.-T. (2008). Galectins: structure, function and therapeutic potential, *Expert Reviews in Molecular Medicine* **10**, e17. A comprehensive account of structures and possible functions of all members of the galectin family.

QUESTIONS

9.1   Discuss the various roles of carbohydrate recognition in the immune system.

9.2   Go to the genomic resource for animal lectins at **http://www.imperial.ac.uk/research/ animallectins/**. In the C-type lectin-like domain (CTLD) database, first look at the domain organization of mammalian proteins containing CTLDs, then examine the sequences of the CTLDs in group II (type II receptors) and group V (NK receptors). Discuss why some of these receptors are not predicted to bind sugars, while others would be expected to bind either mannose or galactose.

References: Drickamer, K. (1992). Engineering galactose-binding activity into a C-type mannose-binding protein, *Nature* **360**, 183–186.

Weis, W.I., Drickamer, K., and Hendrickson, W.A. (1992). Structure of a C-type mannose-binding protein complexed with an oligosaccharide, *Nature* **360**, 127–134.

9.3   Discuss the importance of recognition of sialic acid in biological processes, giving details of the receptors and ligands involved.

9.4   Describe some of the approaches that have been used to characterize glycoprotein ligands for the selectins.

Lead references: Fieger, C.B., Sassetti, C.M., and Rosen, S.D. (2003). Endoglycan, a member of the CD34 family, functions as an L-selectin ligand through modification with tyrosine sulfation and sialyl Lewis[x], *Journal of Biological Chemistry* **278**, 27390–27398.

Hernandez Mir, G., Helin, J., Skarp, K.-P., Cummings, R.D., Makitie, A., Renkonen, R., and Leppanen, A. (2009). Glycoforms of human endothelial CD34 that bind L-selectin carry sulfated sialyl Lewis[x] capped O- and N-glycans, *Blood* **114**, 733–741.

Levinvitz, A., Muhlhoff, J., Isenmann, S., and Westweber, D. (1993). Identification of a glycoprotein ligand for E-selectin on mouse myeloid cells, *Journal of Cell Biology* **121**, 449–459.

9.5   Compare the structures and functions of C-type lectins and galectins.

9.6   No clear function for sialoadhesin has so far been determined although it is thought to be involved in adhesion interactions of macrophages. Discuss how the experiments presented in the following paper provide evidence that sialoadhesin might function as a pathogen receptor.

Reference: Jones, C., Virji, M., and Crocker, P. (2003). Recognition of sialylated meningo-coccal lipopolysaccharide by siglecs expressed on myeloid cells leads to enhanced bacterial uptake, *Molecular Microbiology* **49**, 1213–1225.

Glycoprotein trafficking in cells and organisms

LEARNING OBJECTIVES

By the end of this chapter, students should understand:

1   The roles of calnexin, calreticulin, and M-type lectins in quality control of glycoprotein biosynthesis in the endoplasmic reticulum

2   The functions of mannose 6-phosphate receptors and other lectins in targeting of glycoproteins to specific intracellular locations

3   The mechanisms of selective glycoprotein clearance from circulation

4   Shared features of the sugar-binding mechanisms of different carbohydrate-recognition domain families

Many lectins are integral membrane proteins orientated so that the carbohydrate-recognition domains (CRDs) are located in the extracellular or luminal spaces. This orientation reflects the role of the lectins in binding glycans that have similar dispositions outside the cell and within the luminal compartments. This chapter is devoted to lectins that function in the trafficking of glycoproteins by recognizing endogenous glycans. Intracellular lectins often recognize core sugar structures that are common to many glycoproteins and are present early on in biosynthesis, and they control the flow of the glycoproteins through the secretory pathway. Cell surface lectins often bind to the more variable terminal sugars that are elaborated late in biosynthesis, allowing selective uptake of subsets of glycoproteins from the circulation.

10.1   Lectins have important functions in the secretory pathway

N-linked glycosylation of secretory and cell surface glycoproteins is often portrayed as a terminal and irreversible post-translational modification. However, these glycans are repeatedly modified as glycoproteins move through the secretory pathway. The transient forms of the glycans often serve as handles for sorting events within cells. When the primary function of glycosylation at a particular site

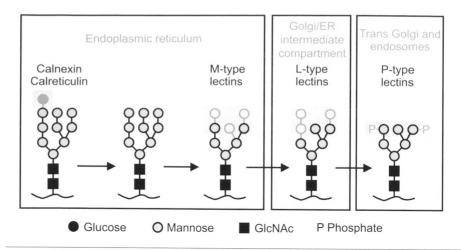

Figure 10.1 Lectins that sort secretory and membrane glycoproteins as they pass through the luminal compartments to the cell surface. Terminal residues that interact wit h these lectins are highlighted in blue. ER, endoplasmic reticulum.

is fulfilled during transit through intracellular compartments, the final glycan may be essentially a by-product of the secretion process. At least four different types of lectins, described in the next few sections, direct intracellular glycoprotein traffic (Figure 10.1).

10.2   Calnexin and calreticulin help glycoproteins fold in the endoplasmic reticulum

Calreticulin, a soluble protein in the lumen of the endoplasmic reticulum, functions in parallel with its membrane-bound homologue calnexin as part of a quality control system that ensures proper folding of proteins destined for the cell surface. Both calnexin and calreticulin bind to the monoglucosylated form of the N-linked oligosaccharide that results from removal of two glucose residues from the glycan initially transferred to the nascent polypeptide.

⊘ See sections 2.4 and 2.5 for more on steps in glycan biosynthesis.

Calnexin and calreticulin are associated with ERp57, a member of the protein disulphide isomerase family (Figure 10.2). ERp57 can assist the folding of luminal glycoproteins that must form disulphide bonds. Removal of the final glucose residue from the N-linked glycan may occur while it is bound to calnexin or calreticulin, causing the glycoprotein to be released. Alternatively, the glucose residue may be trimmed off when the glycoprotein dissociates from calnexin or calreticulin as a result of the equilibrium binding process. Either way, the released, unglucosylated polypeptide follows one of two distinct routes depending on the state of the polypeptide. If the polypeptide is folded correctly, it moves on to the Golgi apparatus. However, if the polypeptide is partially unfolded, the glycoprotein is a substrate

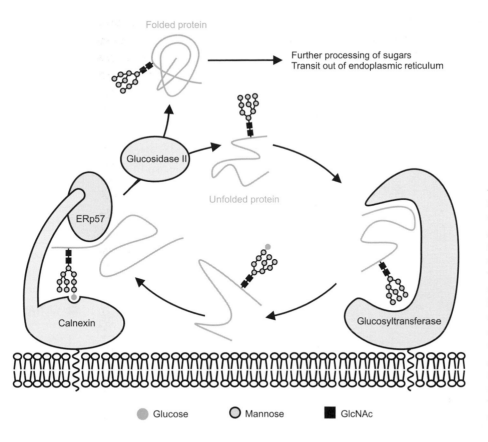

Figure 10.2 The calnexin cycle. Glycoproteins bearing glycans with terminal glucose residues are retained in the endoplasmic reticulum and folding is catalysed by factors such as ERp57. Once glucose is removed by glucosidase II, a correctly folded glycoprotein can exit the endoplasmic reticulum, but the glucosyltransferase re-attaches glucose to misfolded glycoproteins so that they are rebound by calnexin.

for a luminal glucosyltransferase, which re-glucosylates the glycan, making it again a ligand for calnexin or calreticulin. Rebound glycoprotein can then again interact with ERp57 and other folding factors to achieve its correctly folded state when it is next released. Calnexin and calreticulin appear to have very similar functions, although they interact with different but overlapping sets of secretory and membrane glycoproteins.

The overall result of this cycle is that calnexin and calreticulin retain glycoproteins in the endoplasmic reticulum and present them to folding factors such as ERp57 until they have folded correctly. The role of the glucosyltransferase is to distinguish between folded and unfolded states of the protein. In addition to a catalytic domain, this enzyme has a sensor domain that interacts with hydrophobic patches exposed at the surface of partially folded glycoproteins, while the polar surface of a fully folded protein does not allow such interactions. Calnexin and calreticulin each consist of a globular domain, which is folded as a β-sandwich, and an extended arm formed by an antiparallel loop of polypeptide inserted between strands in the globular domain. The sugar-binding site is located in the globular domain while ERp57 binds to the extended arm, which may also interact with the substrate glycoprotein.

10.3   Lectins are involved in degradation of misfolded glycoproteins

Following removal of the glucose residues, the next processing event in the N-linked glycoprotein biosynthetic pathway is the removal of the terminal α1-2-linked mannose residue from the middle branch of the Man$_9$ oligosaccharide to leave Man$_8$. This reaction is catalysed by endoplasmic reticulum mannosidase I (ERManI). For many proteins, this processing is followed by onward transport to the Golgi apparatus where further processing and elaboration of complex glycans takes place (Figure 10.3). However, misfolded glycoproteins encounter a lectin in the membrane of the

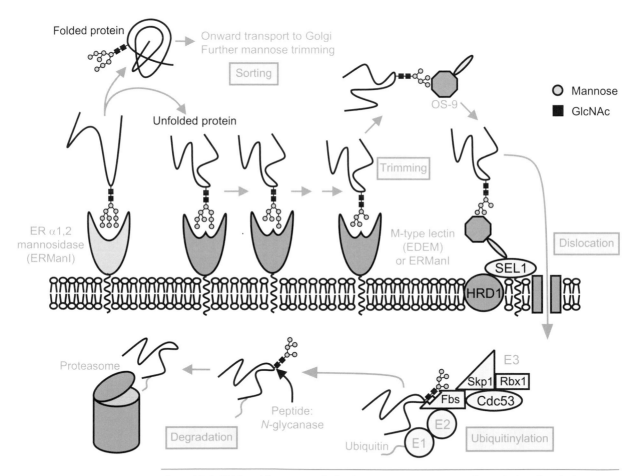

Figure 10.3 ERAD and the proposed role of M-type lectins and OS-9. The endoplasmic reticulum α1-2 mannosidase, ERManI generates a Man$_8$ glycan on folded and unfolded glycoproteins. Trimming of additional mannose residues from the glycans on unfolded glycoproteins by ERManI aided by the M-type lectins (EDEM 1, 2, and 3) leads to recognition by OS-9 and retrotransposition of the misfolded glycoproteins into the cytoplasm. In the cytoplasm, sugar-binding F box proteins form part of the E3 ubiquitin ligase component of the enzyme complex that tags the glycoproteins with ubiquitin to target them for proteasomal degradation.

endoplasmic reticulum, which triggers a process of endoplasmic reticulum-associated protein degradation (ERAD). ERAD involves translocation to the cytoplasm and digestion by proteasomes that also degrade misfolded cytoplasmic proteins. The ERAD pathway is upregulated under stress conditions, such as heat shock, in which the concentration of misfolded proteins increases. Similarly, mutant proteins that fold poorly are particularly good substrates for the degradation machinery.

In mammalian cells, ERAD involves removal of most or all of the α1-2-linked mannose residues. Processing to at least Man_7 by removal of the α1-2-linked mannose on the 1-6 arm of Man_8 is required to trigger ERAD. Thus, slow trimming of mannose residues provides a timer mechanism allowing recognition of misfolded glycoproteins in the endoplasmic reticulum. Targeting of misfolded glycoproteins to ERAD involves at least two types of lectin. The M-type lectins, EDEM 1, 2, and 3 are mannosidase-like proteins that are structurally related to the processing mannosidase, ERManI. A structurally distinct group of lectins including OS-9 contain domains similar in sequence to the P-type carbohydrate-recognition domains of the mannose 6-phosphate receptors (see section 10.6). EDEMs 1, 2, and 3 and OS-9 are all resident in the endoplasmic reticulum and are upregulated under conditions of stress. Overexpression of EDEM1 or EDEM3 enhances de-mannosylation and degradation of misfolded glycoproteins, possibly because EDEM1 and EDEM3 have weak mannosidase activity. However, this activity has not yet been demonstrated with the isolated proteins. OS-9 binds Man_{5-7} glycans with terminal α1-6-linked mannose residues and associates with a membrane protein SEL1 that is part of a larger membrane complex that contains HRD1, an ubiquitin ligase, and a poorly defined membrane-spanning pore for translocation of ERAD substrates.

One model proposed for ERAD is that α1-2-linked mannose residues on glycoproteins that fail to fold to their native conformation are removed by ERManI aided by EDEMs either as co-factors or as active mannosidases. The resulting Man_{5-7} structures are bound by OS-9, which delivers them to the HRD1-SEL1 complex for translocation to the cytoplasm (Figure 10.3). Alternatively, the lectin activity of OS-9 may be required for interaction with glycans on SEL1 rather than on the misfolded glycoproteins. Thus, although the involvement of N-glycans and mannose trimming in degradation of glycoproteins is widely accepted, the exact mechanisms of ERAD are still being uncovered. Questions remain about which enzymes remove the additional mannose residues, which proteins recognize the misfolded glycoproteins, and which proteins are involved in translocation of ERAD substrates back across the endoplasmic reticulum membrane.

Once misfolded glycoproteins have been translocated back into the cytoplasm, they are tagged with ubiquitin to make them substrates for the proteasome. Addition of ubiquitin to lysine residues is catalysed by a series of enzymes. E3 ubiquitin ligases catalyse the transfer of activated ubiquitin, after first recognizing the target protein and recruiting an E1 ubiquitin activating enzyme and an E2 ubiquitin conjugating enzyme. There are many different types of E3 ubiquitin ligases to allow recognition of a wide variety of protein substrates. One type of E3 ligase specifically targets glyco-proteins through recognition of N-linked glycans. Fbs1, specific to neuronal cells, and Fbs2, which is more widely expressed, are sugar-binding F-box proteins that are found in a group of E3 ligases called SCF complexes. Each SCF complex consists of three common protein components (Skp1, Cul1, and Rbx1) and an F-box protein that determines the substrate specificity. F-box proteins bind to Skp1 through an F-box domain and to the protein substrate through various types of associated domains. Fbs1 and Fbs2 initiate ubiquitination of glycoproteins after binding to the two GlcNAc residues in the core of N-linked glycans. Following addition of ubiquitin to a glycoprotein, the glycans are removed by a peptide:N-glycanase, before degradation of the deglycosylated polypeptide in the proteasome.

10.4   L-type lectins transport glycoproteins from the endoplasmic reticulum to the Golgi

The endoplasmic reticulum-Golgi intermediate compartment (ERGIC) has been defined by the identification of proteins that can be found in the endoplasmic reticu-lum but which are concentrated in a distinct region of the cell. One of the best char-acterized of these proteins is a membrane-bound lectin, ERGIC-53. ERGIC-53 interacts with glycoproteins bearing mannose-containing glycans, showing broad specificity for high-mannose oligosaccharides of the type found on proteins moving from the endoplasmic reticulum to the Golgi. Because ERGIC-53 cycles between the endoplasmic reticulum and the ERGIC, it acts as a cargo receptor that mediates transport of glycoproteins between these compartments (Figure 10.4).

Glycobiology and disease
Blood clotting deficiency

The role of ERGIC-53 in the transport of certain specific glycoproteins is demon-strated by the effect of mutations in the human *ERGIC-53* gene that result in a clotting deficiency caused by decreased serum levels of coagulation factors V and VIII. In the absence of functional ERGIC-53, transport of these highly glycosylated coagulation factors through the secretory pathway is very slow, resulting in the decreased serum levels. The absence of a generalized glycoprotein secretion defect in individuals with the affected genotype indicates that ERGIC-53 is required only for trafficking of a spe-cific subset of glycoproteins.

A second lectin, VIP36 (vesicular integral membrane protein of 36 kDa), also binds mannose-containing glycoproteins and is found in the ERGIC. VIP36 cycles between the ERGIC and the *cis*-Golgi. It is tempting to speculate that VIP36 is involved in moving glycoproteins between these two compartments, but this role has not been extensively investigated. Both ERGIC-53 and VIP36 have luminal domains that are structurally related to the legume lectins (see Chapter 11) and are thus desig-nated L-type lectins. A third L-type lectin, VIP36-like protein (VIPL) is resident in the endoplasmic reticulum and may bind to folded glycoproteins immediately after their release from the calnexin–calreticulin cycle.

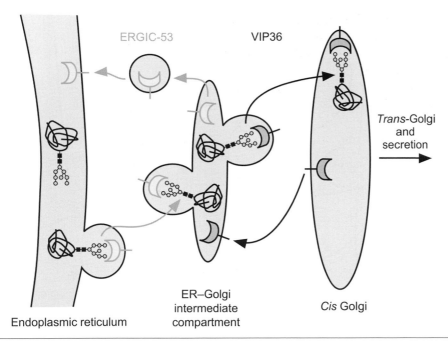

Figure 10.4 Trafficking patterns for the L-type lectins ERGIC-53 and VIP36. The lectins mediate transport of selected glycoproteins bearing high mannose oligosaccharides into and out of the endoplasmic reticulum (ER)–Golgi intermediate compartment.

10.5   Mannose 6-phosphate residues target lysosomal enzymes to lysosomes

Lysosomal proteases, nucleases, glycosidases, and other hydrolases catalyse degradation of molecules that are sent to lysosomes for destruction. Lysosomal hydrolases are synthesized together with secretory glycoproteins in the rough endoplasmic reticulum. To prevent harmful enzymes from being released from the cell, an effective method of targeting newly synthesized hydrolases to the lysosomes is required. This process is mediated by **mannose 6-phosphate residues** on the oligosaccharide chains of enzymes destined for the lysosomes.

Upon translocation into the lumen of the endoplasmic reticulum, the hydrolases acquire $Glc_3Man_9GlcNAc_2$ oligosaccharide chains that are processed in the normal way until they reach the Golgi apparatus. In the *cis*-Golgi, the oligosaccharides of the hydrolases are modified in a different way from those of secretory glycoproteins by the addition of one or more mannose 6-phosphate residues. The mannose 6-phosphate residues are then bound by mannose 6-phosphate receptors when the lysosomal enzymes reach the *trans*-Golgi network (Figure 10.5). The receptor–enzyme complexes are transported, via clathrin-coated vesicles, to late endosomes, where the enzymes are released from the receptors due to acidic conditions. The enzymes are then packaged into lysosomes, where the phosphate tag is removed,

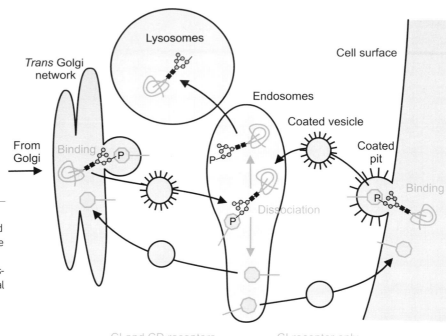

Figure 10.5 Routes of glycoprotein trafficking mediated by the two mannose 6-phosphate receptors. Both receptors can mediate transport from the *trans*-Golgi apparatus to the endosomal sorting compartments, but only the cation-independent receptor mediates uptake from the extracellular medium.

Glycobiology and disease
I-cell disease

while the receptors recycle back to the *trans*-Golgi network to pick up more enzymes. In addition, mannose 6-phosphate receptors can cycle to the plasma membrane to re-capture hydrolases that escape through the default secretory pathway.

Studies of patients with I-cell disease, a lysosomal storage disorder, have provided key insights into the mannose 6-phosphate recognition system. Lysosomes from I-cell disease sufferers contain large amounts of undigested material because lysosomal hydrolases normally found in lysosomes are missing. High levels of lysosomal enzymes are found in the blood of these patients, indicating that the enzymes are largely secreted rather than being targeted efficiently to lysosomes. The oligosaccharide chains of newly synthesized lysosomal hydrolases of I-cell disease patients do not contain mannose 6-phosphate residues due to a deficiency of the first of two enzymes needed to create this modification (Figure 10.6). This phosphotransferase adds a GlcNAc-phosphate residue to the 6-OH group of mannose residues. A phosphodiesterase then removes the GlcNAc residue to leave mannose 6-phosphate. Because only enzymes destined for lysosomes are tagged with mannose 6-phosphate, the phosphotransferase must recognize both the acceptor oligosaccharide and the protein to which it is attached. The enzyme is highly specific for lysosomal enzymes to prevent mis-targeting of secretory proteins to the lysosomes. It is not yet clear exactly how this specificity is achieved although lysine residues on the surface of the lysosomal enzymes are required for recognition by the GlcNAc-phosphotransferase.

10.6 Two types of mannose 6-phosphate receptor take part in lysosomal enzyme targeting

Two structurally distinct mannose 6-phosphate receptors bind with high affinity to oligosaccharides bearing terminal mannose 6-phosphate residues (Figure 10.6). The smaller cation-dependent mannose 6-phosphate receptor (CD-MPR) binds mannose 6-phosphate in the presence of divalent cations such as Ca^{2+} or Mn^{2+}. The larger cation-independent mannose 6-phosphate receptor (CI-MPR) does not require cations for ligand binding. Both receptors bind ligand most efficiently at pH 6–7, allowing them to interact with hydrolases in the *trans*-Golgi network, and they release the lysosomal enzymes in the more acidic environment of the endosomes.

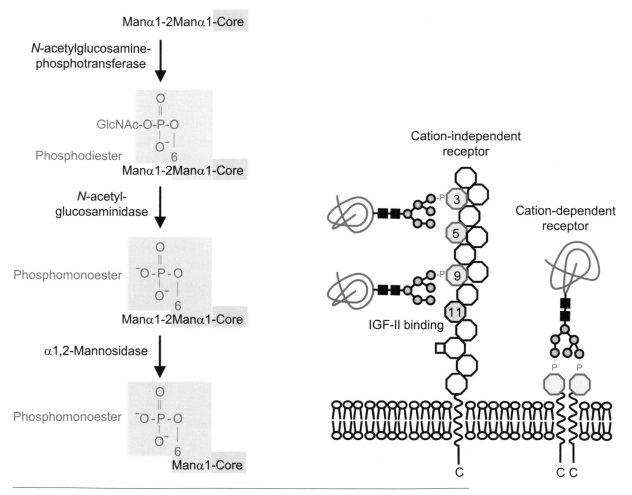

Figure 10.6 Biosynthesis of the mannose 6-phosphate recognition marker on the termini of high-mannose N-linked glycans and summary of the structures of the two mannose 6-phosphate receptors. CRDs that bind ligands are shaded. IGF-II, insulin-like growth factor II.

Gene knockout experiments show that both mannose 6-phosphate receptors are required for the efficient targeting of lysosomal enzymes. Mutant mice lacking the CD-MPR have increased levels of phosphorylated lysosomal enzymes in their serum, indicating that the CI-MPR cannot fully compensate for the loss of CD-MPR. Fibroblasts lacking the CI-MPR have reduced levels of lysosomal enzymes as well as increased accumulation of undigested material in lysosomes. The change in lysosome function is more dramatic in fibroblasts lacking CI-MPR than in those without CD-MPR. Analysis of mice lacking the CI-MPR is complicated by the fact that this receptor also functions in clearance of insulin-like growth factor II. Insulin-like growth factor II binds to the CI-MPR at the cell surface and is targeted to the lysosomes for degradation, although it does not bear mannose 6-phosphate residues and binds to the receptor at a site different to that used by the lysosomal enzymes.

The need for two mannose 6-phosphate receptors for the efficient targeting of lysosomal enzymes may be due to the fact that they have slightly different specificities and each binds a subset of lysosomal enzymes. Each receptor recognizes mannose 6-phosphate residues on the enzymes, but the number and position of mannose 6-phosphate residues can alter the affinity for each of the receptors. For example, the CD-MPR binds more tightly to oligosaccharides with two phosphorylated mannose residues than to those with only one, whereas the CI-MPR binds equally well to glycans with one or two mannose 6-phosphate residues. The CI-MPR, but not the CD-MPR, can also bind glycans that have been modified by the addition of GlcNAc-1-phosphate by the phosphotransferase, but which have not been acted on by the phosphodiesterase, so that a phosphodiester remains. This activity of the CI-MPR allows lysosomal enzymes to be targeted correctly, even if they escape the action of the phosphodiesterase in the *cis*-Golgi.

The differences in affinities displayed by the two receptors are likely to be due to the different arrangement of binding sites in the receptor polypeptides (Figure 10.6). The mannose 6-phosphate-binding domains in the receptors are known as P-type CRDs. The luminal region of the CD-MPR consists of a single P-type CRD that binds one mannose 6-phosphate residue. However, the receptor dimerizes so that two mannose 6-phosphate residues can be bound. The CI-MPR contains 15 P-type CRD-like domains, but only three of these domains bind mannose 6-phosphate. Domains 3 and 9 bind mannose 6-phosphate with high affinity while domain 5 has only low affinity for mannose 6-phosphate but binds tightly to GlcNAc-1-P-Man phosphodiesters. The combination of three different binding sites for phosphorylated sugars in the CI-MPR allows high affinity binding to a wide range of phosphorylated glycans. Thus, the CI-MPR and the dimeric CD-MPR bind preferentially to hydrolases with different numbers of glycans that are singly or doubly phosphorylated.

Key interactions with mannose 6-phosphate in the binding site of P-type CRDs take the form of hydrogen bonds with the characteristic hydroxyl groups at positions 2, 3, and 4, making the site selective for mannose (Figure 10.7). Conserved arginine

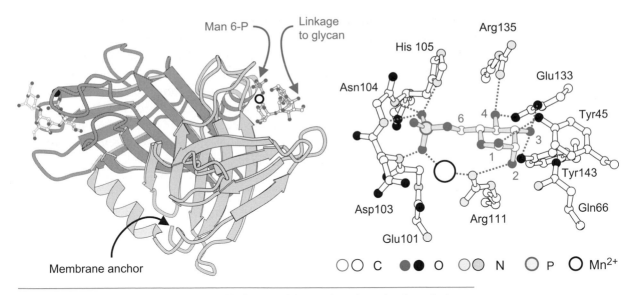

Figure 10.7 Overall fold and monosaccharide-binding site of the CRD from the cation-dependent mannose 6-phosphate receptor. [Based on entry 1C39 in the Protein Databank.]

residues characteristic of P-type CRDs are involved in these hydrogen bonds and do not interact with the negatively charged phosphate substituent. Like the binding site in the C-type CRDs, the P-type CRD ligand-binding site of the CD-MPR involves a divalent cation. However, in this instance the bound Mn^{2+} interacts with the phosphate portion of the ligand rather than with hydroxyl groups on the sugar. Most of the remaining interactions of the protein with the phosphate are through hydrogen bonds with backbone atoms.

The mannose 6-phosphate receptors recognize a slightly modified form of the oligosaccharide that is added to all glycoproteins early in the secretory pathway. The ability to modify and then to recognize a subset of these high mannose oligosaccharides in the secretory pathway is a recent event in evolutionary terms, because the use of mannose 6-phosphate receptors for intracellular enzyme targeting is restricted to cells of higher animals. Invertebrates do not have functional mannose 6-phosphate receptors. In this respect, the mannose 6-phosphate targeting system differs from the systems described in preceding sections, in which recognition of sugars added to all glycoproteins early in the secretory pathway plays fundamental roles in all eukaryotic cells.

10.7   The asialoglycoprotein receptor clears altered serum glycoproteins into the liver

As well as mediating trafficking between cellular compartments, lectins are also involved in transport of glycoproteins into cells. Both the **asialoglycoprotein receptor** and the **mannose receptor** bind circulating glycoproteins and mediate their endocytosis. The asialoglycoprotein receptor binds glycans containing terminal galactose or *N*-acetylgalactosamine (GalNAc) residues. Glycoproteins bearing such glycans can be created by sialidase (neuraminidase) treatment of commonly occurring serum glycoproteins to remove terminal sialic acid residues. The desialylated glycoproteins have a very short half-life in serum compared with the untreated, fully sialylated forms (Figure 10.8). Glycoproteins bound by the asialoglycoprotein receptor at the surface of hepatocytes are internalized through clathrin-coated pits. Following uncoupling of the ligand–receptor complex in endosomes, the glycoproteins are directed to lysosomes for degradation (Figure 10.9). The asialoglycoprotein receptor was the first animal lectin discovered. It is a trimer of subunits, each of which contains a galactose-binding C-type CRD. The trimer binds with high affinity to tri-antennary N-linked glycans.

It has been proposed that the asialoglycoprotein receptor might have a general role in the clearance of serum glycoproteins. A sialidase that removes sialic acid residues from complex oligosaccharides as glycoproteins circulate could gradually uncover terminal galactose residues. Once enough galactose residues become exposed, the glycoprotein would be bound by the asialoglycoprotein receptor and removed from the circulation. Such a mechanism would allow the half-life of a serum glycoprotein to be controlled by the number of complex oligosaccharides that it bears, and their accessibility to sialidase. In this way, one receptor could control the half-life of many different serum glycoproteins. However, mice lacking functional asialoglycoprotein receptor do not show a defect in serum glycoprotein turnover and they do not accumulate endogenous asialoglycoproteins, even though they do not clear galactose-terminated glycoproteins injected into the circulation. In addition, no sialidase that could remove sialic acid from glycoproteins in the circulation has yet been identified. Therefore, it seems more likely that the receptor has a specialized function in clearing glycoproteins with terminal galactose or GalNAc residues that are minor components of serum, though physiological ligands with these terminal sugars remain to be identified.

An alternative role for the receptor in regulating the levels of serum glycoproteins containing glycans with terminal sialic acid linked α2-6 to GalNAcβ1-4GlcNAc or to Galβ1-4GlcNAc has also been proposed. Although capped with sialic acid, the GalNAc or galactose residues on these glycoproteins can fit into the binding sites of the asialoglycoprotein receptor because the α2-6 linkage of sialic acid leaves the 3- and 4-hydroxyl groups free. Thus, only sialylated glycans with sialic acid linked α2-3 to galactose or GalNAc need to be desialylated to allow binding to the receptor. Very few glycoproteins contain glycans with sialic acid linked α2-6 to GalNAcβ1-4GlcNAc. Examples include the pregnancy-specific glycoprotein glycodelin in humans and prolactin-like hormones in rat. In contrast, oligosaccharides with sialic acid linked α2-6 to Galβ1-4GlcNAc are common on circulating glycoproteins. Because these glycoproteins are so abundant in blood, it is hard to demonstrate rapid clearance of proteins with terminal NeuAc α2-6Gal because the asialoglycoprotein receptors on the liver would normally be saturated with these ligands. However, levels of some common serum glycoproteins bearing glycans with terminal NeuAcα2-6Gal,

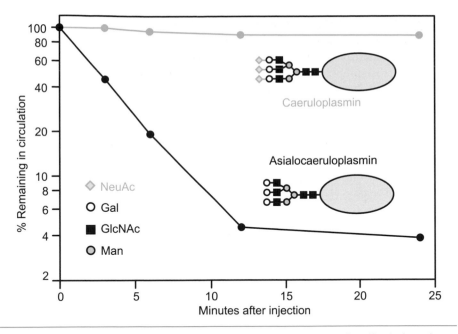

Figure 10.8 Kinetics of clearance of asialoglycoproteins from circulation mediated by the hepatic asialoglycoprotein receptor. Radioactively labelled serum caeruloplasmin that has been treated with sialidase is cleared rapidly from circulation. [Adapted from: Morell, A.G., Irvine, R.A., Sternlieb, I., Scheinberg, I.H., and Ashwell, G. (1968). Physical and chemical studies on ceruloplasmin: V. Metabolic studies on sialic-free ceruloplasmin *in vivo, Journal of Biological Chemistry* **243**, 155–159.]

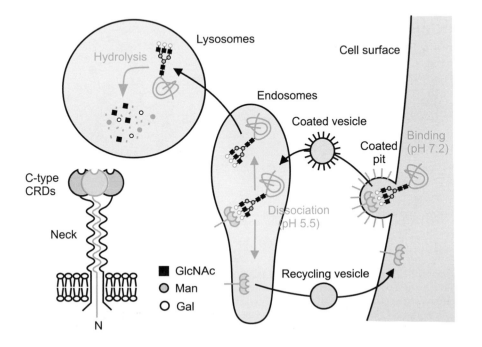

Figure 10.9 Interaction of the asialoglycoprotein receptor with terminal galactose residues on a desialylated glycoprotein, and the pathways of uptake and degradation of asialoglycoproteins and receptor recycling. The receptor contains two types of subunits that bind preferentially to different branches of an N-linked glycan.

including haptoglobin and serum amyloid protein, are raised in the serum of the asialoglycoprotein receptor knockout mice, providing evidence that clearance by the receptor may contribute to regulation of the turnover of these proteins.

The galactose-binding C-type CRD of the asialoglycoprotein receptor has an almost identical fold to the CRD of MBP. The C-type CRDs discussed in Chapter 9 bind hexoses such as GlcNAc and mannose, while the asialoglycoprotein receptor binds hexoses related to galactose. The distinction between these two groups of sugars is in the orientation of the 4-hydroxyl group. Comparison of the sequences of the mannose- and galactose-binding CRDs reveals an important difference in the Ca^{2+}-ligating residue in the binding site (Figure 10.10). Binding to mannose-type sugars would require rotation of the 3-hydroxyl group into an unfavourable conformation. In contrast, binding of the 3- and 4-hydroxyl groups of galactose or GalNAc can occur with the preferred positioning of the hydroxyl groups. Substitution of the asialoglycoprotein receptor residues into MBP results in a change in specificity from mannose to galactose. Incorporation of a tryptophan residue adjacent to the galactose-binding site in the modified CRD of MBP leads to high-affinity galactose binding comparable with that of the asialoglycoprotein receptor, due to a hydrophobic packing interaction between the aromatic ring of tryptophan and the non-polar B face of galactose. Selectivity for galactose over mannose requires insertion of a glycine-rich loop from the asialoglycoprotein receptor, which prevents the tryptophan

⊙ See section 7.2 for more on the polarity of sugar residues.

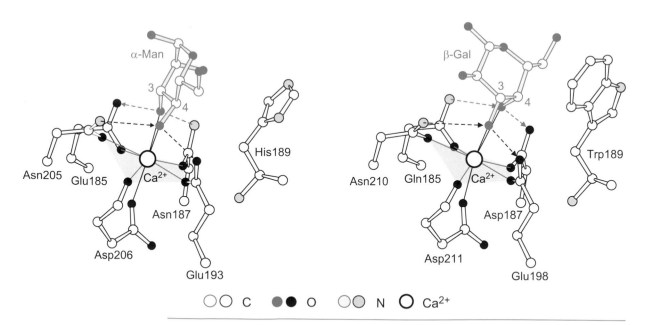

Figure 10.10 Modification of MBP to bind galactose. The native binding site in MBP is shown on the left and the modifications introduced to generate a site that mimics the asialoglycoprotein receptor are shown on the right. The residues that interact with the 3-hydroxyl group of mannose are glutamic acid and asparagine, while these residues are glutamine and aspartic acid in the galactose-binding site. [Based on entries 2MSB and 1AFA in the Protein Databank.]

Box 10.1   185

BOX 10.1 Glycobiology of disease: *The Asialoglycoprotein receptor may help to prevent sepsis*

Endogenous sialidases that could desialylate glycans on circulating glycoproteins to generate ligands for the asialoglycoprotein receptor have not been identified. However, many bacterial pathogens express sialidases, so it is possible that desialylated ligands for the receptor are produced during bacterial infections. Studies in mice infected with *Streptococcus pneumoniae*, used as a model for human blood poisoning resulting from severe bacterial infection (sepsis), show that glycoproteins on platelets become desialylated by the bacterial enzyme neuraminidase A. The infected mice have a decreased platelet count in the blood (thrombocytopenia) due to clearance of desialylated platelets by the asialoglycoprotein receptor. In asialoglycoprotein receptor-deficient mice infected with *Streptococcus pneumoniae*, platelets are desialylated to the same extent as in wild-type infected mice, but they are not cleared from the blood so platelet levels remain normal. The asialoglycoprotein receptor-deficient mice died sooner after infection than the wild-type infected mice, and showed evidence of more severe effects on the blood coagulation system including increased formation of blood clots in the vessels of the kidney and liver, and haemorrhage of the spleen.

These findings are of interest clinically because patients with sepsis due to infection with *Streptococcus pneumoniae* or other bacteria often develop a life-threatening condition, disseminated intravascular coagulation (DIC). In DIC, there is both excessive clotting in small blood vessels and enhanced bleeding due to disruption of the blood coagulation system. In the mouse studies of sepsis, DIC-like symptoms of excessive clotting and bleeding were more evident in the asialoglycoprotein-receptor deficient mice, suggesting that clearance of desialylated platelets by the asialoglycoprotein receptor can delay the severe coagulation defect that leads to death. Lowering the number of platelets available could slow clotting and prevent the consumption of clotting factors that leads to excessive bleeding. The situation in humans may be more complex; whereas a reduction in platelet numbers appears to be protective in the infected mice, lowered platelet levels in patients with sepsis are associated with higher mortality. In addition, DIC can also occur in sepsis caused by bacteria that do not express sialidases so clearance of platelets by the asialoglycoprotein receptor would not occur in these cases. Nevertheless, these experiments do highlight the fact that, although desialylated ligands for the asialoglycoprotein receptor may not be found under normal physiological conditions, they can be produced under pathological conditions.

Essay topic:

• Discuss how the asialoglycoprotein receptor might contribute to the control of haemostasis.

Lead references:

• Ellies, L.G., Ditto, D., Levy, G.G., Wahrenbrock, M, Ginsburg, D., Varki, A., Le, D.T., and Marth, J.D. (2002). Sialyltransferase ST3Gal-IV operates as a dominant modifier of hemostasis by concealing asialoglycoprotein receptor ligands, *Proceedings of the National Academy of Sciences U.S.A.* **99**, 10042–10047.

• Grewal, P.K., Uchiyama, S., Ditto, D., Varki, N., Dzung, D.T., Nizet, V., and Marth, J.D. (2008). The Ashwell receptor mitigates the lethal coagulopathy of sepsis, *Nature Medicine* **14**, 648–655.

• Rumjantseva, V., Grewal, P.K., Wandall, H.H., Joseffson, E.C., Sorensen, A.L., Larson, G., Marth, J.D., Hartwig, J.H., and Hoffmeister, K.M. (2009). Dual roles for hepatic lectin receptors in the clearance of chilled platelets, *Nature Medicine* **15**, 1273–1280.

side chain from moving. When fixed in position, this side chain prevents mannose from binding, while in the absence of the glycine-rich loop, the indole ring can move out of the way and allow mannose to enter the binding site. Thus, specificity for galactose over mannose results from two effects: accommodation of the galactose ligand by making favourable hydrogen bonds and hydrophobic interactions, and exclusion of mannose by creating a steric clash. These results illustrate that, because binding sites in the C-type CRDs are shallow indentations on the surface of the domain and specificity for different classes of sugars is based on relatively few contacts between protein and sugar ligand, the selectivity for sugars can be changed by making only a small number of amino acid changes.

Because of the disposition of the 3- and 4-hydoxyl groups in galactose, the galactose ring in the binding site is turned roughly 90° compared with the way mannose binds. As a result of this re-orientation, the 2-position of galactose comes much closer to the surface of the protein than does the 2-position of mannose. The extended 2-substituent of GalNAc is able to interact with the surface of the CRD, allowing the asialoglycoprotein receptor to bind preferentially to GalNAc compared with galactose.

10.8   The mannose receptor removes naturally occurring glycoproteins from circulation

Intracellular glycoproteins released from cells in response to pathological events display high-mannose oligosaccharides that are recognized by the mannose receptor (Figure 10.11). These glycoproteins, including lysosomal enzymes, tissue plasminogen activator, and neutrophil myeloperoxidase are secreted into the blood or the extravascular spaces during inflammation, infection, and tissue damage. The released lysosomal enzymes and myeloperoxidase play an important part in attacking invading pathogens, whereas tissue plasminogen activator initiates the process of dissolution of blood clots. Because these enzymes have potentially harmful activities, their levels in the circulation must be tightly regulated so that they do not cause damage. Recognition of the high-mannose oligosaccharides on these proteins by the mannose receptor in the liver, or on macrophages, leads to their internalization and degradation. The mannose receptor is the main mechanism for clearance of lysosomal enzymes and myeloperoxidase, but tissue plasminogen

◔ See section 8.7 for more on tissue plasminogen activator.

activator is cleared in parallel by the low-density lipoprotein receptor-related protein on hepatocytes.

Mice in which the mannose receptor gene has been knocked out do not clear mannose-terminated glycoproteins into the liver, and there are increased levels of several lysosomal enzymes in their circulation. In addition to its role in removing products of the inflammatory response, the mannose receptor binds to the propeptide of type I procollagen. This peptide, which bears a single high-mannose oligosaccharide, is cleaved from procollagen during the formation of collagen fibrils in tissue re-modelling and wound healing. Levels of circulating collagen propeptide are also elevated in the mannose receptor knockout mice.

10.9   The mannose receptor also regulates activity of sulphated hormones

A surge in the concentration of the pituitary hormone lutropin triggers ovulation, and an episodic rise and fall in the circulating levels of this hormone is essential for efficient activation of its receptor on target cells. This rise and fall is achieved through regulated release of lutropin from the pituitary, followed by rapid clearance from the circulation once it has acted on the target cells. Lutropin bears unusual complex oligosaccharides that terminate in 4-O-sulphated GalNAc residues (Figure 10.12). Recognition of these GalNAc 4-SO_4 residues by the mannose receptor in the liver results in fast removal of lutropin from the circulation. Sulphated GalNAc residues are bound by the N-terminal domain of the mannose receptor (Figure 10.11). This domain is homologous to the plant lectin ricin, and is called an R-type CRD. Thus, the mannose receptor is unusual among lectins in that it contains two different types of CRD that recognize sugars by different mechanisms.

Lutropin and follitropin, a pituitary hormone that stimulates development of the endometrium during the first phase of the female menstrual cycle, have identical α subunits. The two hormones also have homologous β subunits. During biosynthesis, a protein-specific GalNAc-transferase adds GalNAc to lutropin glycans and a GalNAc 4-sulphotransferase, GalNAc-4-ST1, adds sulphate to the resulting GalNAcβ1-4GlcNAcβ1-2Man structures. These enzymes are both highly expressed in the pituitary gland. The protein-specific GalNAc-transferase recognizes a cluster of basic amino acid residues in the α subunit of lutropin. This enzyme recognition sequence is present in the identical α subunit of follitropin, but in this hormone, the sequence is not accessible to the enzyme. Thus, galactose and sialic acid residues are added to the glycans on follitropin to form standard bi-antennary, complex oligosaccharides (Figure 10.12). As a consequence, this hormone has a long half-life in the circulation. The GalNAc 4-SO_4 modification also appears on thyrotropin, which is another, structurally related glycoprotein hormone produced in the pituitary, as well as on a few other unrelated glycoproteins that contain the peptide recognition motif for the GalNAc-transferase. Addition of the GalNAc 4-SO_4 recognition tag to

➲ See section 8.6 for more on pituitary hormones.

➲ See section 11.5 for more on R-type CRDs.

R-type CRD
SO4-4GalNAcβ1-4GlcNAα
Pituitary hormones
Fibonectin
type II repeat
1
2
3
C-type CRDs
4
5
6
Manα1
6
3
Manα1
6
3
Manβ1-4GlcNAcβ1-4GlcNAcβ1-Asn
Manα1
2
Manα1
Manα1
Released lysosomal enzymes
7
8
C

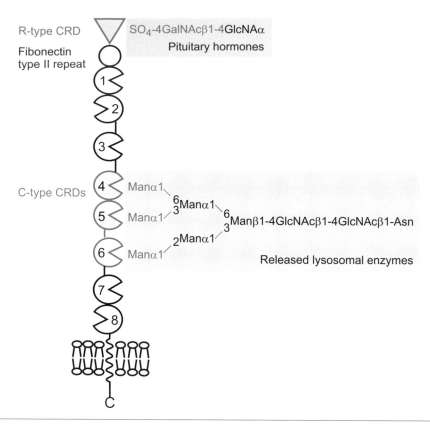

Figure 10.11 Overall organization of the mannose receptor, highlighting its extended arrangement of different types of protein modules that interact with ligands containing high mannose oligosaccharides and sulphated ligands.

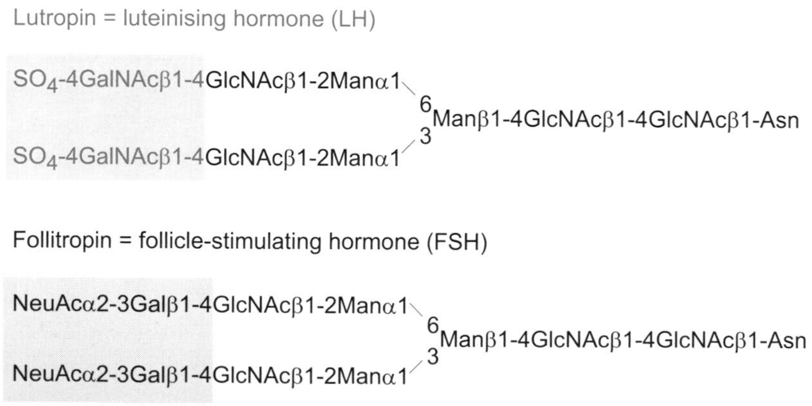

Lutropin = luteinising hormone (LH)
SO4-4GalNAcβ1-4GlcNAcβ1-2Manα1
6
Manβ1-4GlcNAcβ1-4GlcNAcβ1-Asn
3
SO4-4GalNAcβ1-4GlcNAcβ1-2Manα1
Follitropin = follicle-stimulating hormone (FSH)
NeuAcα2-3Galβ1-4GlcNAcβ1-2Manα1
6
Manβ1-4GlcNAcβ1-4GlcNAcβ1-Asn
3
NeuAcα2-3Galβ1-4GlcNAcβ1-2Manα1

Figure 10.12 N-linked glycans attached to pituitary glycoprotein hormones. Hormones that are rapidly removed from circulation are capped with GalNAc 4-SO_4 residues rather than the terminal sialic acid residues typical of longer-lived serum glycoproteins.

Box 10.2   189

BOX 10.2 Glycotherapeutics: *Glycosylation of recombinant glycoproteins must be carefully controlled*

The use of recombinant DNA technology has allowed the production of human proteins that could not easily be purified from natural sources in sufficient quantities to make their therapeutic use practical. For example, recombinant human insulin for treatment of diabetes is now produced by expression in the bacterium *Escherichia coli*.

Bacteria or yeast are often the hosts of choice for expressing proteins for use in research, because they allow production of large amounts of protein rapidly and relatively cheaply. However, recombinant human proteins for therapeutic use must usually be produced in mammalian cells to ensure correct post-translational modification, particularly glycosylation (see Chapter 13). Glycosylation of recombinant proteins needs to be carefully controlled because incorrect glycosylation can dramatically affect the half-life of a protein in the circulation, as well as sometimes affecting the activity of proteins directly.

The hormone erythropoietin is a good example of a human glycoprotein that has successfully been produced in recombinant form for therapeutic use. Erythropoietin released into the blood from the kidney stimulates growth and maturation of red blood cells from precursor cells in the bone marrow. A recombinant form of erythropoietin, Epoetin alfa, produced in Chinese hamster ovary (CHO) cells has been in clinical use for treatment of anaemia since 1988. It is mainly used to treat anaemia caused by kidney failure, when production of natural erythropoietin is reduced, or anaemia that results as a side-effect of chemotherapy treatment for cancer. Erythropoietin has three N-linked glycosylation sites and one O-linked glycosylation site. Comparison of the glycosylation of erythropoietin isolated from human urine with that of the recombinant form shows that, for this glycoprotein, CHO cells can produce correct glycosylation. In both natural and recombinant erythropoietin, all three N-linked sites are occupied mostly with sialylated complex tetra-antennary oligosaccharides. The main difference between the two forms is that the oligosaccharides of natural erythropoietin contain both α2-3- and α2-6-linked sialic acid, whereas only α2-3-linked sialic acid is found on recombinant erythropoietin because CHO cells do not express the sialyl α2-6 transferase.

The difference in sialic acid linkage does not affect the activity of erythropoietin, but the extent of sialylation of the oligosaccharides does affect its survival in the circulation. Erythropoietin treated with a sialidase to remove sialic acid has a greatly reduced activity *in vivo* due to rapid clearance from the circulation via the asialoglycoprotein receptor. Mutagenesis experiments to remove or introduce glycosylation sites show that, even without sialidase treatment, survival of erythropoietin in the circulation is linked to the number of oligosaccharides present on the polypeptide, with more highly glycosylated forms lasting longer. This effect of glycosylation has been exploited to produce a longer lasting recombinant form of erythropoietin. Darbepoetin alfa has five N-linked oligosaccharides as a result of site-directed mutagenesis to introduce two additional Asn-X-Thr sequences into the erythropoietin polypeptide. Addition of the two extra oligosaccharides does not affect the structure or activity of erythropoietin, but this recombinant form survives three times longer in the circulation than recombinant erythropoietin with three N-linked oligosaccharides. Like Epoetin alfa, Darbepoetin alfa has proved to be a safe and effective treatment for anaemia. Because it has longer *in vivo* activity, Darbepoetin alfa can be administered less frequently than Epoetin alfa, meaning fewer injections for patients.

As well as being used clinically to treat anaemia, recombinant erythropoietin is misused by some athletes. Stimulation of red blood cell production can be advantageous in endurance sports such as cycling because with more red blood cells, more oxygen is transported to the muscles. Use of erythropoietin is banned in sport and much effort has been directed towards developing tests that can distinguish between natural and recombinant forms in human urine, so that doping can be detected. Here, the close similarity of Epoetin alfa to natural erythropoietin is a disadvantage, but techniques such as iso-electric focusing or two-dimensional gel electrophoresis in combination with immunoblotting can distinguish the two forms due to the very minor differences in glycosylation.

Essay topic

- Discuss why careful analysis of the glycosylation of recombinant glycoproteins is important when developing forms of these proteins for therapeutic use.

Lead references

- Elliott, S., Lorenzini, T., Asher, S., Aoki, K., Brankow, D., Buck, L., Busse, L., Chang, D., Fuller, J., Grant, J., Hernday, N., Hokum, M., Hu, S., Knudten, A., Levin, N., Komorowski, R., Martin, F., Navarro, R., Osslund, T., Rogers, G., Rogers, N., Trail, G., and Egrie, J. (2003). Enhancement of therapeutic protein activities through glycoengineering, *Nature Biotechnology* **21**, 414–421.
- Elliott, S., Egrie, J., Browne, J., Lorenzini, T., Busse, L., Rogers, N., and Ponting, I. (2004). Control of rHuEPO biological activity: The role of carbohydrate, *Experimental Hematology* **32**, 1146–1155.
- Takeuchi, M., Takasaki, S., Miyazaki, H., Kato, T., Hoshi, S., Kochibe, N., and Kobata, A. (1988). Comparative study of the asparagine-linked sugar chains of human erythropoietins purified from urine and the culture medium of recombinant Chinese hamster ovary cells, *Journal of Biological Chemistry* **263**, 3657–3663.

selected glycoproteins is similar to the addition of mannose 6-phosphate residues to lysosomal enzymes, because the transferases that initiate addition of these two modifications must each recognize structural features of the glycoproteins to which the specific tags are to be added, as well as the acceptor oligosaccharides

The importance of the specific GalNAc 4-SO_4 recognition tag on the oligosaccharides of lutropin for the physiological function of this hormone is highlighted by the phenotype of mice in which the gene for the GalNAc 4-sulphotransferase, GalNAc-4-ST1, is knocked out. Female mice lacking the gene reach sexual maturity more quickly than wild-type mice, have a longer oestrus cycle, and produce more litters. In the absence of the sulphotransferase, GalNAc residues are added to the oligosaccharides on lutropin but they are not sulphated and so cannot be recognized by the R-type CRD of the mannose receptor. As a result, the half-life of lutropin in the circulation is increased allowing more stimulation of oestrogen production. Interestingly it appears that in the knockout mice, most of the oligosaccharides of lutropin are sialylated, with sialic acid linked α2-6 to GalNAc. Lutropin with oligosaccharides bearing terminal NeuAcα2-6 GalNAc may bind to the asialoglycoprotein receptor, accounting for the finding that although the half-life of lutropin is longer in the knockout mice it is still cleared from the circulation more quickly than if it was merely being filtered through the kidneys.

10.10   Some intracellular lectins have roles in the nucleus

At least two galectins, galectin 1 and galectin 3, are transported into the nucleus. The distribution of galectin 3 between the cytoplasm and the nucleus varies depending on the proliferative state of the cell. In both human and mouse cells, galectin 3 is present mainly in the nucleus when cells are dividing, but is predominantly in the cytoplasm in quiescent cells. Both galectin 1 and galectin 3 co-localize in the nucleus with components of the pre-mRNA splicing machinery, and galectin 3 is found in high molecular weight complexes containing ribonucleo proteins.

Good biochemical evidence for a role for both of these proteins in RNA splicing has been provided using *in vitro* splicing assays. For example, nuclear extracts of tumour cells from which galectin 1 or 3 have been removed by affinity chromatography are deficient in splicing activity and the activity can be reconstituted by the addition of either recombinant galectin 1 or recombinant galectin 3. However, knockout mice lacking either or both of galectins 1 and 3 are viable, suggesting that any role of these proteins in splicing is not essential. It is unlikely that the participation of galectins in the splicing process involves carbohydrate recognition. Although many nuclear proteins are modified by the addition of O-linked GlcNAc (see Chapter 3), this sugar is not a ligand for the galectins, and galactose residues are not known to be present on cytoplasmic or nuclear proteins. In addition, although saccharide ligands can inhibit the splicing activity of galectins 1 and 3 in the *in vitro* assays, mutation of a critical asparagine residue in the CRD of galectin 1 abolishes sugar binding but has no effect on splicing activity. Thus, it is possible that the galectin CRD mediates a non-lectin, protein–protein interaction that is involved in splicing.

SUMMARY

The intracellular recognition events discussed in the first part of this chapter involve glycans common to all proteins that are N-glycosylated. Glycans of this type tag many different glycoproteins that serve a variety of different functions at the cell surface. Similarly, the asialoglycoprotein and mannose receptor clearance systems can handle a multitude of different glycoproteins as long as they bear the correct glycans. However, two of the systems discussed, the mannose 6-phosphate receptor and the binding of glycans containing 4-O-sulphated GalNAc residues to the mannose receptor, involve unusual glycans that are attached to specific subsets of glycoproteins. These special glycans allow targeting of selected glycoproteins in ways that are different from the bulk of glycoproteins. Attachment of special glycans to the selected subset of glycoprotein targets is directed by information in the amino acid sequence of the glycoprotein. Because the amino acid sequence is encoded in a gene, this arrangement provides a mechanism by which targeting information can be encoded in the gene.

KEY REFERENCES

Aebi, M., Bernasconi, R., Clerc, S., and Molinari, M. (2009). N-glycan structures: recognition and processing in the ER, *Trends in Biochemical Sciences* **35**, 74–82. This review, and the one by Lederkremer (2009) (see below), provide critical analysis of our current understanding of the roles of glycans and proteins that recognize them in glycoprotein folding, quality control, and transport.

Dagher, S. F., Wang, J. L., and Patterson, R. J. (1995). Identification of galectin-3 as a factor in pre-mRNA splicing, *Proceedings of the National Academy of Sciences U.S.A.* **92**, 1213–1217. This paper, and the one by Voss *et al.* (2008) (see below), present evidence for the involvement of galectins in RNA splicing.

Fiete, D. J., Beranek, M. C., and Baenziger, J. U. (1998). A cysteine-rich domain of the 'mannose' receptor mediates GalNAc-4-SO$_4$ binding, *Proceedings of the National Academy of Sciences U.S.A.* **95**, 2089–2093. Experiments showing that the R-type CRD of the mannose receptor binds to the sulphated sugars on glycoprotein hormones are described.

Kamiya, Y., Kamiya, D., Yamamoto, K., Nyfeler, B., Hauri, H.-P., and Kato, K. (2008). Molecular basis of sugar recognition by the human L-type lectins ERGIC-53, VIPL, and VIP36, *Journal of Biological Chemistry* **283**, 1857–1861. Experiments to define the precise glycan-binding specificities of three L-type lectins are presented.

Kato, K. and Kamiya, Y. (2007). Structural views of glycoprotein-fate determination in cells, *Glycobiology* **17**, 1031–1044. An overview of the structures and mechanisms of action of the enzymes and lectins involved in glycoprotein folding, quality control, and transport.

Kim, J.-J. P., Olson, L.J., and Dahms, N.M. (2009). Carbohydrate recognition by the mannose-6-phosphate receptors, *Current Opinion in Structural Biology* **19**, 534–542. A detailed review of the structure and functions of mannose 6-phosphate receptors.

Kolatkar, A.R., Leung, A.K., Isecke, R., Brossmer, R., Drickamer, K., and Weis, W.I. (1998). Mechanism of *N*-acetylgalactosamine binding to a C-type animal lectin carbohydrate-recognition domain, *Journal of Biological Chemistry* **273**, 19502–19508. The structural basis for selective binding of GalNAc and galactose to a C-type CRD, like that of the asialoglycoprotein receptor, is described.

Lederkremer, G.Z. (2009). Glycoprotein folding, quality control and ER-associated degradation, *Current Opinion in Structural Biology* **19**, 515–523. See Aebi *et al.* (2009) above.

Lee, S. J., Evers, S., Roeder, D., Parlow, A. F., Risteli, J., Risteli, L., Lee, Y. C., Feizi, T., Langen, H., and Nussenzweig, M. C. (2002). Mannose receptor-mediated regulation of serum glycoprotein homeostasis, *Science* **295**, 1898–1901. This paper presents analysis of the mannose receptor knockout mouse, providing evidence for the receptor's role in glycoprotein clearance.

Mi, Y., Fiete, D., and Baenziger, J.U. (2008). Ablation of GalNAc-4-sulfotransferase-1 enhances reproduction by altering the carbohydrate structures of luteinizing hormone in mice, *Journal of Clinical Investigation* **118**, 1815–1824. The phenotype of mice deficient in the enzyme responsible for sulphation of GalNAc residues on lutropin is analysed.

Mikami, K., Yamaguchi, D., Tateno, H., Hu, D., Qin, S.-Y., Kawasaki, N., Hiarbayashi, J., Ito, Y., and Yamamoto, K. (2010). The sugar-binding ability of human OS-9 and its involvement in ER-associated degradation, *Glycobiology* **20**, 310–321. Experiments defining the specificity of OS-9 for high mannose oligosaccharides and providing evidence for a role of OS-9 in ERAD are presented.

Roberts, D.L., Weix, D.J., Dahms, N.M., and Kim, J.-J.P. (1998). Molecular basis of lysosomal enzyme recognition: three-dimensional structure of the cation-dependent mannose 6-phosphate receptor, *Cell* **93**, 639–648. The crystal structure of a P-type CRD is described.

Schrag, J. D., Bergeron, J. J. M., Li, Y., Borisova, S., Hahn, M., Thomas, D. Y., and Cygler, M. (2001). The structure of calnexin, an ER chaperone involved in quality control of protein folding, *Molecular Cell* **8**, 633–644. The crystal structure of calnexin is described, giving insights into the mechanism of oligosaccharide binding.

Steirer, L.M., Park, E.I., Townsend, R.R., and Baenziger, J.U. (2009). The asialoglycoprotein receptor regulates levels of plasma glycoproteins terminating with sialic acid α2,6-galactose, *Journal of Biological Chemistry* **284**, 3777–3783. Experiments showing that some sialylated glycoproteins could be ligands for the asialoglycoprotein receptor are described.

Tozawa, R., Ishibashi, S., Osuga, J., Yamamoto, K., Yagyu, H., Ohashi, K., Tamura, Y., Yahagi, N., Iizuka, Y., Okazaki, H., Harada, K., Gotoda, T., Shimano, H., Kimura, S., Nagai, R., and Yamada, N. (2001). Asialoglycoprotein receptor deficiency in mice lacking the major receptor subunit: its obligate requirement for the stable expression of oligomeric receptor, *Journal of Biological Chemistry* **276**, 12624–12628. Analysis of glycoprotein clearance in the asialoglycoprotein receptor knockout mouse is presented.

Voss, P.G., Gray, R.M., Dickey, S.W., Wang, W., Park, J.W., Kasai, K., Hirabayashi, J., Patterson, R.J., and Wang, J.L. (2008). Dissociation of the carbohydrate-binding and splicing activities of galectin-1, *Archives of Biochemistry and Biophysics* **478**, 18–25. See Dagher *et al.* (1995) above.

QUESTIONS

10.1 What effects might you expect treatment of a cell line with either castanospermine, an inhibitor of glucosidase I, or kifunensine, an inhibitor of ER mannosidase, to have on the synthesis and secretion of glycoproteins?

10.2 What are the advantages of using oligosaccharides rather than amino acid sequences as recognition markers in glycoprotein folding, trafficking, and clearance?

10.3 Discuss the functions of lectins in the secretory pathway.

10.4 Describe how the different properties of multiple mannose 6-phosphate binding sites in the cation-independent mannose 6-phosphate receptor have been characterized.

References: Bohnsack, R.N., Song, X., Olson, L.J., Kudo, M., Gotschall, R.R., Canfield, W.M., Cummings, R.D., Smith, D.F., and Dahms, N.M. (2009). Cation-independent mannose 6-phosphate receptor. A composite of distinct phosphomannosyl binding sites, *Journal of Biological Chemistry* **284**, 35215–35226.

Hancock, M.K., Yammani, R.D., and Dahms N.M. (2002). Localization of the carbohydrate recognition sites of the insulin-like growth factor II/mannose 6-phosphate receptor to domains 3 and 9 of the extracytoplasmic region, *Journal of Biological Chemistry* **277**, 47205–47212.

Hancock, M.K., Haskins, D.J., Sun, G., and Dahms, N.M. (2002). Identification of residues essential for carbohydrate recognition by the insulin-like growth factor II/mannose 6-phosphate receptor, *Journal of Biological Chemistry* **277**, 11255–11264.

Westlund, B., Dahms, N.M., and Kornfeld, S. (1991). The bovine mannose 6-phosphate/insulin-like growth factor II receptor. Localization of mannose 6-phosphate binding sites to domains 1-3 and 7-11 of the extracytoplasmic region, *Journal of Biological Chemistry* **266**, 23233–23239.

10.5   Discuss the experimental and medical evidence that ERGIC-53 is needed to transport some glycoproteins out of the endoplasmic reticulum.

References: Appenzeller, C., Andersson, H., Kappeler, F., and Hauri H.-P. (1999). The lectin ERGIC-53 is a cargo transport receptor for glycoproteins, *Nature Cell Biology* **1**, 330–334.

Moussali, M., Pipe, S.W., Hauri, H.P., Nichols, W.C., Ginsburg, D., and Kaufman, R.J. (1999). Mannose-dependent ERGIC-53-mediated ER to Golgi trafficking of coagulation factors V and VIII, *Journal of Biological Chemistry* **274,** 32539–32542.

Nichols, W.C., Seligsohn, U., Zivelin, A., Terry, V.H., Hertel, C.E., Wheatley, M.A., Moussali, M.J., Hauri, H.-P., Ciavarella, N., Kaufman, R.J., and Ginsburg, D. (1998). Mutations in the gene for ERGIC-53, a protein of the endoplasmic reticulum-Golgi intermediate compartment, cause combined deficiency of coagulation factors V and VIII, *Cell* **93**, 61–70.

10.6   Discuss how experiments with knockout mice have provided strong evidence that the mannose receptor clears mannose-terminated glycoproteins from the circulation, but have cast doubt on a role for the asialoglycoprotein receptor in clearance of desialylated glycoproteins. Refer to the papers by Lee *et al.* (2002) and Tozawa *et al.* (2001) in the reference list.

Glycobiology of plants, bacteria, and viruses

LEARNING OBJECTIVES

By the end of this chapter, students should understand:

1 The use of glycan derivatives as signalling molecules in plants

2 The mechanisms of plant and bacterial toxin action

3 Exploitation of cell surface glycans as receptors for bacteria and viruses

Although the major focus of this book is on mammalian glycobiology, some interesting aspects of the glycobiology of plants and micro-organisms are considered in this chapter. Comparative glycobiology provides useful clues to the evolution of glycans and their functions. The last part of the chapter discusses how the functions of glycans and lectins have changed through evolution.

11.1 Plant and microbial sugars have some functions not seen in mammals

Some functions of glycans are conserved throughout evolution, whereas others are more specialized adaptations of mammalian cells. Equally, plants and micro-organisms use glycans in some ways that are not seen in mammalian cells. Although plants, fungi, and some types of bacteria synthesize glycoproteins with N-linked glycans, these glycans often differ from mammalian oligosaccharides. For example, fungal oligosaccharides contain mainly mannose residues, while plants incorporate xylose into complex oligosaccharides (Figure 11.1). Also, unlike mammalian cells, plant and fungal cells are enclosed within polysaccharide-containing cell walls. A further, unique aspect to the glycobiology of plants is that they use free oligosaccharides as signalling molecules.

Plants, viruses, and bacteria all make use of carbohydrate-recognition proteins. Plant lectins are some of the best understood examples of carbohydrate-recognition systems and they continue to serve as models for how selective binding to oligosaccharides can

➲ See section 9.3 for more on fungal glycans.

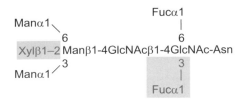

Figure 11.1 Example of a plant N-linked glycan that differs from mammalian N-linked structures. The core α1–3-linked fucose residue and the xylose residue are not found in mammalian glycans.

be achieved. They are also very powerful tools for detecting glycans. Plant and bacterial toxins, as well as bacterial and viral sugar-binding adhesion molecules, all bind to glycans on the surfaces of mammalian cells. They represent the converse of the interaction between the surfaces of micro-organisms and mannose-binding protein and other parts of the mammalian innate immune system.

11.2   Some bacteria have protein glycosylation pathways that are related to the mammalian glycosylation machinery

The existence of N- and O-linked protein glycosylation in prokaryotes has been known for many years, but the full complement of structures and their similarity to eukaryotic glycosylation has only recently become apparent. Of particular interest has been the finding that *Campylobacter jejuni* synthesizes glycoproteins in which sugars are transferred from a lipid-linked intermediate to glycoproteins at Asn-X-Ser/Thr sequence, suggesting a close parallel with glycosylation in the endoplasmic reticulum of eukaryotic cells (Figure 11.2). Remarkably, all of the genes required for this complex glycosylation are found in a single cluster, which can be transferred to *Escherichia coli* and confers the ability to N-glycosylate proteins. Although the details of both the glycans transferred and the target sequence differ from their eukaryotic counterparts, recent studies indicate that it is possible to modify the gene cluster so that typical mammalian core glycans are transferred in the bacteria. In addition, related systems in other bacterial species have different requirements. Thus, with further engineering, it will probably be possible fully to mimic the eukaryotic system, opening up the possibility of producing therapeutic glycoproteins in bacteria.

⊙ See **section 13.5** for more on therapeutic glycoproteins.

In addition to the N-linked glycosylation systems, O-linked glycosylation systems are known in *Campylobacter* and other eubacterial species. In addition, both N- and O-linked glycosylation systems have been described in *Archaebacteria*, so these types of glycosylation are found in all three kingdoms of life. Although only distantly related, key enzymes in the glycosylation pathway, such as oligosaccharyltransferase, show sequence similarity indicating that these pathways share a common evolutionary origin.

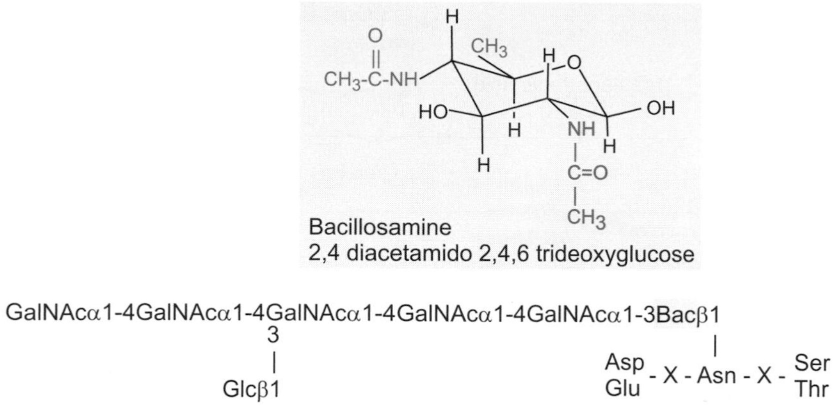

Bacillosamine
2,4 diacetamido 2,4,6 trideoxyglucose

GalNAcα1-4GalNAcα1-4GalNAcα1-4GalNAcα1-4GalNAcα1-3Bacβ1
                          3 |
                          | Asp - X - Asn - X - Ser
                       Glcβ1   Glu   Thr

Figure 11.2 N-linked glycosylation in *Campylobacter jejuni*. An example of a glycan attached through the novel sugar bacillosamine is shown, along with the consensus sequence of attachment of sugars to asparagine residues in *Campylobacter*.

11.3   Plants use oligosaccharides as signalling molecules

Oligosaccharides that function as signalling molecules in plants are often called oligosaccharins. Oligosaccharins that regulate growth and development or defence reactions are released from the cell walls of micro-organisms, or from the cell walls of the plants themselves. For example, oligogalacturonides are oligomers of α-1-4-linked galacturonic acid (Figure 11.3) that are released from polysaccharides of the plant cell wall in response to wounding and infection. Mechanical disruption of the cell wall during wounding can generate oligogalacturonides, but a wound-inducible polygalacturonase enzyme also specifically catalyses their release. They can also be produced by the action of polygalacturonases secreted by attacking fungi or bacteria. Released oligogalacturonides stimulate transcription of defence genes that produce responses to protect the plant against both pathogens and herbivores. These responses include enhanced ion flux across the plasma membrane and production of reactive oxygen species. Similar defence responses are induced when chitinases in plant tissues break down chitin in the cell walls of invading pathogenic fungi to form soluble oligomers of β1-4-linked GlcNAc.

Oligosaccharide derivatives also act as signalling molecules that initiate the symbiotic relationship between leguminous plants (peas, beans, and related species) and nitrogen-fixing bacteria. Rhizobial bacteria live in root nodules, where they reduce atmospheric nitrogen to ammonia that can be used by the plant. The symbiotic relationship between legumes and *Rhizobia* is specific. Each type of legume can only be infected by a limited number of rhizobial species. The specificity is conferred by *N*-acylated chitin oligomers that are called Nod factors because they direct formation of the root nodules (Figure 11.3). Nod factors are synthesized by rhizobial bacteria in response to the release of flavenoids from the roots of the plant.

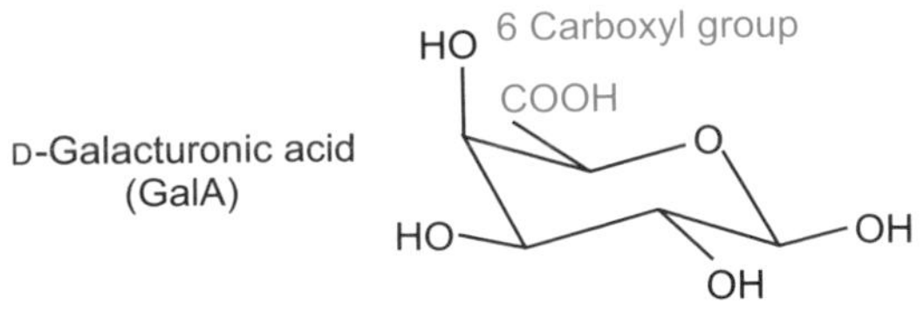

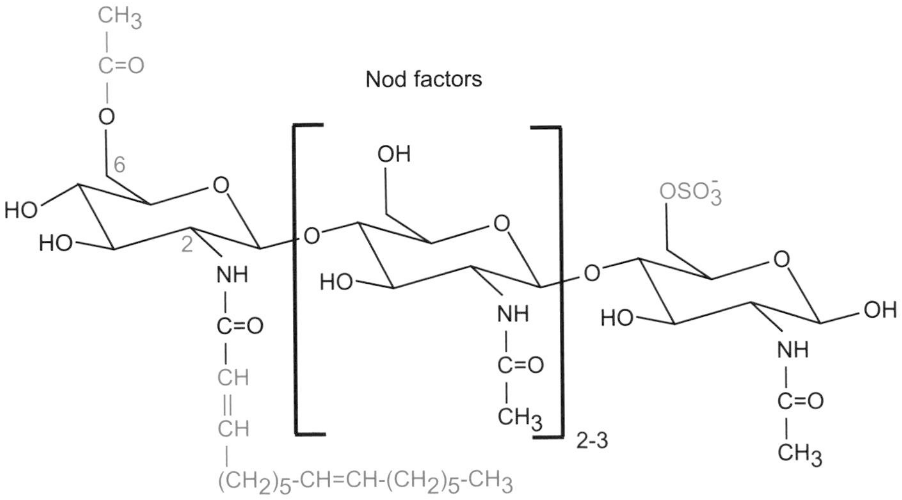

Figure 11.3 Oligosaccharide-based signalling molecules in plants. Polymers of α1-4-linked galacturonic acid are released from cell walls and induce defence responses. Nod factors are derivatives of chitin, a polymer of β1-4-linked GlcNAc, that are produced by rhizobial bacteria to induce developmental changes in root hairs. Parts of the Nod factors that vary between different species of *Rhizobia* are indicated in blue.

Nod factors play two parts in the formation and infection of nodules. First, they induce a deformation of the plant root hairs, which initiates the infection process. Second, the Nod factors activate division of cells within the root cortex leading to the development of nodules that are invaded by bacteria from the primary infection site. Different rhizobial species produce Nod factors that differ in the extent of modification of the basic oligosaccharide structure.

Like hormones in mammals, the plant signalling oligosaccharides are active at very low concentrations, so their receptors must bind them with high affinity. Detection of the Nod factors by the plants must involve Nod-specific receptors in the plant roots. Potential Nod factor receptors have been identified by genetic analysis of mutant leguminous plants that are unable to produce root nodules. These membrane bound proteins have extracellular regions containing lysin motif (LysM) domains that are similar to the peptidoglycan-binding domains found in many bacterial and eukaryotic proteins and could bind Nod factors, and intracellular serine/threonine kinase domains for signal transduction. However, direct binding of Nod factors to these receptors has not yet been demonstrated. In the model plant, *Arabidopsis*, similar receptors containing intracellular kinase domains and extracellular LysM domains that bind chitin oligosaccharides are essential for eliciting defence reactions against fungal pathogens. The many well-characterized plant lectins discussed in the next section, which are mainly found in bulbs and seeds, do not take part in these signalling processes.

11.4   Common plant lectins are useful tools for biologists

Lectins in seeds and fruits defend plants against predators and pests. Toxic reactions induced by lectins cause gastrointestinal distress, which can condition animals to avoid eating particular plants. Lectins in legumes also combat invasion of seeds by microbial pathogens. Plant lectins have been known and studied for far longer than animal lectins.

Lectins from a family of plants often form a group of related proteins, although some may bind mannose-containing oligosaccharides while other members of the same group bind galactose-containing structures. Lectins in each group are composed of polypeptides with a characteristic fold. Most plant lectins are oligomers and all are multivalent. The constituent polypeptides can have more than one binding site for sugar, leading to the formation of large clusters of binding sites in the oligomers. This high degree of multivalency explains the efficient haemagglutination mediated by these proteins, which serves as a convenient way to assay their activity (Figure 11.4). The position of binding sites at opposite ends of lectin oligomers, similar to the arrangement in dimeric galectins, explains the ability to bridge between cell surfaces in the haemagglutination assay.

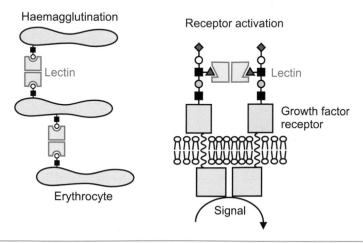

Figure 11.4 Role of plant lectin multivalency in haemagglutination and receptor activation. Cross-linking of glycans on the surfaces of different cells leads to formation of large aggregates of cells, while lateral cross-linking of glycans on two receptor molecules on a single cell can bring their intracellular domains into proximity, leading to activation.

The most extensively studied plant lectins are those found in the seeds of leguminous plants. The first lectin described, concanavalin A (ConA), was isolated from jack beans. Concanavalin A binds to mannose-containing oligosaccharides and has a particular affinity for branched trimannose structures such as those found in the core of N-linked glycans. Because there are several such branched structures in high-mannose oligosaccharides, concanavalin A binds these structures with particularly high affinity.

The subunits of legume lectins have a common overall structure which is stabilized by the presence of two divalent cations near the sugar-binding site (Figure 11.5). This fold is conserved in the L-type animal lectins, although legume lectins are soluble proteins without membrane anchors. In spite of the similarity of their tertiary structure, the arrangements of subunits in legume lectins vary. They are always at least dimeric, accounting for their ability to cross-link glycoproteins, but there is often further association to form tetramers. Subtle differences near the sugar-binding sites account for a range of different ligand-binding specificities. For example, some legume lectins bind with high affinity to the unsialylated core 1 structure of O-linked glycans (Galβ1-3GalNAcβ1-Ser/Thr), a structure that is found

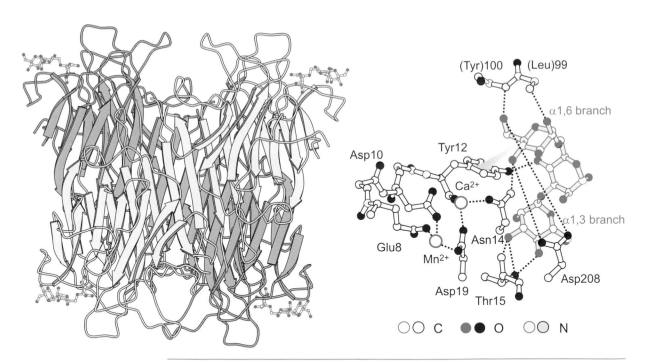

Figure 11.5 Overall structure of concanavalin A and details of the binding site with a trimannose core structure bound. Concanavalin A exists as a tetramer formed as a dimer of dimers. Each of the four identical subunits contains a sugar-binding site. Hydrogen bonds formed between protein residues and hydroxyl groups of mannose residues in the bound trimannose structure are indicated, as well as a hydrophobic packing interaction between Tyr12 and the B face of one mannose residue. The divalent cations stabilize positions of residues involved in sugar ligation, but do not make direct interactions with the sugar. [Based on entry 1CVN in the Protein Databank using Molscript.]

on tumour-associated mucins. The same principles of accommodation and exclusion that account for the binding of different ligands to the C-type animal lectins are also observed for the legume lectins.

● See section 13.8 for more on mucins on tumours.

Much of the interest in concanavalin A and other plant lectins resides in their ability to stimulate quiescent cells to begin DNA synthesis and undergo mitosis. This mitogenic activity results from cross-linking of growth factor receptors through their glycan moieties, mimicking the effects of their natural protein ligands (Figure 11.4). The ready availability of an agent able to stimulate a variety of cells has been very useful in studies of the control of cell growth. The ability of plant lectins to stimulate cell surface receptors also accounts for toxicity associated with eating uncooked beans of many plant species, because the undenatured lectins interact with intestinal epithelial cells.

In addition to being useful because of their activities at the cellular level, the plant lectins are biochemical tools. For example, affinity chromatography on immobilized lectins provides a means of purifying glycoproteins and the oligosaccharides derived from them. The ability of lectins to bind to specific oligosaccharides also makes them useful tools for detecting glycans in cells and tissues. An advantage of this application of the plant lectins is that they can be used to detect specific glycans on intact glycoproteins either in a gel blotting format or on tissues. Their specificity also makes lectins useful diagnostic reagents. For example, lectins that distinguish A and B blood group substances can be used as blood-typing reagents.

● See section 6.9 for more on lectins in cytochemistry

11.5   Some plant lectins are toxins

One group of plant lectins consists of toxins, such as **ricin** from castor beans. The ability of ricin and its homologues to bind sugars provides a mechanism for delivering toxins to the interior of cells. The A or active subunits of the toxins are glycohydrolases that remove a critical base from the RNA component of ribosomes. The resulting inhibition of protein synthesis leads to cytotoxicity (Figure 11.6). Sugar-binding activity resides in the B or binding subunits, which bind to terminal galactose residues on glycoproteins at the surface of animal cells. When some of these proteins move from the cell surface to the interior of the cell during endocytosis, the toxin travels with them. Once in the luminal compartment, the toxin is routed backwards to the endoplasmic reticulum by again binding to terminal galactose residues. In this case, the galactose residues may be on calreticulin, which moves from the Golgi apparatus back to the endoplasmic reticulum, carrying the toxin with it. In the endoplasmic reticulum, the toxin is recognized as a target for cytoplasmic protein degradation and is therefore transported through channels in the endoplasmic reticulum membrane to the cytoplasm. Once in the cytoplasm, it escapes degradation because of its low content of lysine residues and has access to the ribosomes.

The B-subunit of ricin contains two homologous sugar-binding domains (numbered 1 and 2), each folded in a β-trefoil structure in which the three lobes (designated

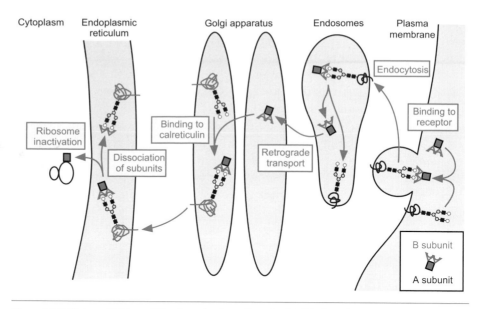

Figure 11.6 Ricin trafficking within a target eukaryotic cell. Initial binding of ricin to cell surface glycoproteins leads to internalization into endosomes. Retrograde transport through the Golgi apparatus to the endoplasmic reticulum depends, at least in part, on binding to galactose-terminated glycoproteins such as calreticulin that cycle between these compartments. In the endoplasmic reticulum, the subunits dissociate and the A subunit is recognized as a misfolded protein and is translocated to the cytoplasm where it inactivates ribosomes.

α, β, and γ) have been formed by an early gene duplication event (Figure 11.7). Although the lobes are homologous, only some contain active sugar-binding sites. Of the total of six lobes in the two domains, three appear to bind sugars (1α, 1β, and 2γ). Each of these lobes contains a key aromatic residue that packs against the B face of galactose in the binding site. The presence of these three sugar-binding sites facilitates high-affinity binding to multivalent galactose-containing ligands.

Ricin-like carbohydrate-recognition domains (R-type CRDs) are found in a diverse set of proteins (Figure 11.6). In bacteria, R-type CRDs are associated with hydrolytic domains, including glycosidases. These CRDs anchor the hydrolases to polysaccharide substrates. Many of the GalNAc transferases that initiate synthesis of O-linked glycans in eukaryotic cells also contain R-type CRDs in addition to catalytic domains. These domains recognize GalNAc residues already attached to serine and threonine residues, and target adjacent residues for further glycosylation. Finally, a ricin-like domain mediates binding of sulphated glycoprotein hormones to the mammalian mannose receptor.

See sections 3.4 and 10.9 for more on R-type CRDs.

11.6   Many bacterial toxins are lectins

Like ricin, many bacterial toxins have a two-subunit organization, although neither subunit shows any sequence similarity to ricin. Bacterial toxins consisting of a single A subunit and five B subunits are designated AB$_5$ toxins. AB$_5$ toxins include:

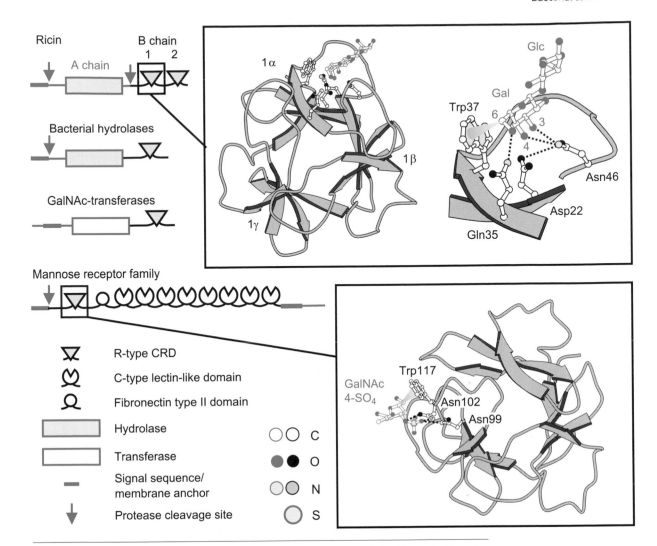

Figure 11.7 Organization of some proteins with ricin-type CRDs, structure of the ricin B subunit, and structure of the R-type CRD from the mannose receptor. Details of the ricin-binding site show a hydrophobic packing interaction between Trp37 and the B face of galactose. Residues that form hydrogen bonds to hydroxyl groups of galactose are also indicated. In the mannose receptor R-type CRD, a hydrophobic interaction between Trp117 and the B face of the GalNAc ring of GalNAc 4-SO$_4$, as well as hydrogen bonds between two asparagine side chains and the sulphate group are indicated. [Based on entries 2AAI and 1DQO in the Protein Databank.]

- cholera toxin, which causes cholera
- *Escherichia coli* enterotoxin, which is responsible for many cases of food poisoning
- pertussis toxin, the causative agent of whooping cough
- Shiga toxin, which causes dysentery
- Shiga-like verotoxins from *E. coli*.

The A subunits of the Shiga and verotoxins inhibit protein synthesis in a similar way to the A subunit of ricin. The A subunits of the remaining toxins have adenosine diphosphate-ribosylating activity, which transfers the adenosine diphosphate and ribose portions of nicotinamide adenine dinucleotide to specific amino acids in heterotrimeric G proteins. The resulting stimulation or interference with signalling cascades leads to effects such as increased water permeability in the gut and hence to diarrhoea (Figure 11.8).

In each of the AB_5 bacterial toxins, the single A subunit is associated with a pentamer of B subunits arranged in a doughnut-like shape. The B subunits bind to cell surface glycans, usually glycolipids (Figure 11.8). Although many of the toxin B subunits are related to each other in sequence, each toxin binds to a somewhat different type of cell surface glycan. The folds of the B subunits of the bacterial toxins are different to those of plant and animal lectins, but the principles of sugar recognition are similar. Key interactions in the binding sites involve hydrophobic packing against the B face of galactose and GalNAc residues, and the *N*-acetyl group of

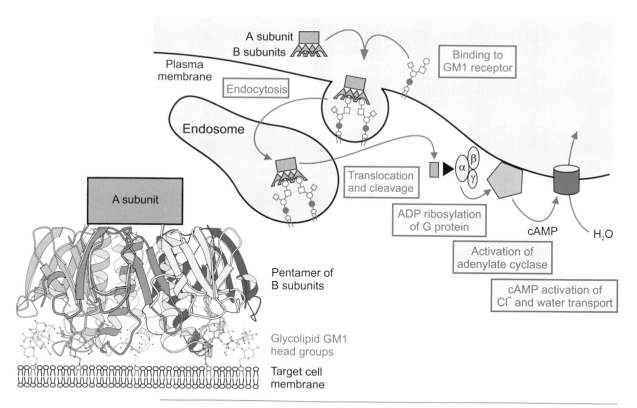

Figure 11.8 Mechanism of action of cholera toxin. Each B subunit interacts with the pentasaccharide head group of a separate glycolipid molecule (GM1) in the target cell membrane, leading to internalization and transfer of the A subunit to the cytoplasm. [Adapted from: Weis, W.I. and Drickamer, K. (1996). Structural basis of lectin-carbohydrate recognition, *Annual Review of Biochemistry* **65**, 441–473. Based on entry 2CHB in the Protein Databank.]

GlcNAc as well as hydrogen bonding to key sugar hydroxyl groups. The binding sites are generally extended and each subunit can interact with as many as five sugar residues. Binding affinity for cell surfaces is significantly enhanced by the oligomeric structure, because binding sites in the five subunits all project from one face of the oligomers (Figure 11.8). High-affinity inhibitors of toxin binding have been created by synthesizing clusters of sugar residues designed to interact multivalently with multiple toxin subunits.

11.7   Bacteria use lectins to bind to host cell surfaces

In addition to producing toxins that bind to cell surface glycans, many types of bacteria adhere to sugar-containing structures on host cell surfaces. Such adhesion can be part of an infection process leading to disease or it can be a normal mechanism of coexistence. For instance, some strains of *E. coli* adhere to intestinal epithelium and form part of the normal gut microflora, while other strains adhere to the epithelium of the urinary tract and are pathogenic. In both types of *E. coli*, the sugar-binding sites are located at the tips of pili that project from the surface of the bacteria (Figure 11.9).

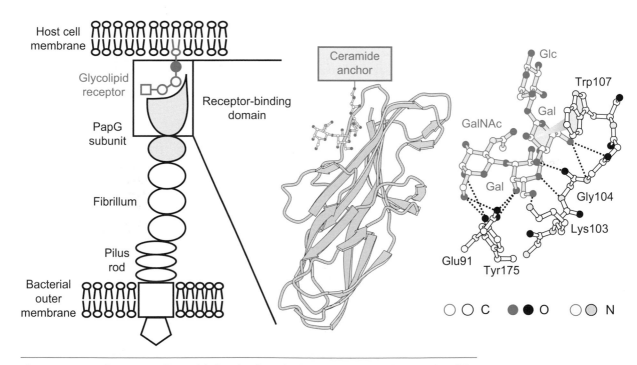

Figure 11.9 Overall structure of bacterial pilus showing subunit arrangement and the structure of the receptor-binding domain of the PapG subunit bound to a glycolipid head group oligosaccharide. The pilus is assembled from subunits that have incomplete immunoglobulin-like folds. The β-sheet structure of each subunit is completed by insertion of a strand from the next subunit along in the pilus. [Based on entry 1J8R in the Protein Databank.]

The pili consist of an inner rod-like structure and a more flexible outer segment. At the tip, a special subunit contains a sugar-binding domain with a unique extended β-barrel structure.

Two different types of pilus in pathogenic *E. coli* have been studied in detail. The adhesin at the tips of type 1 pili is designated FimH and binds to mannose-containing structures. The adhesin on P pili is PapG, one form of which binds a globoside head group that contains GalNAc and galactose residues (GalNAcβ1-3Galβ1-4Galβ1-4Glcβ1-ceramide). This interaction underlies the binding of *E. coli* to kidney epithelium during renal pathogenesis. The structure of the PapG sugar-binding domain is different to the sugar-binding domains in bacterial toxins and in animal lectins, but it utilizes the same principles for sugar recognition. The extended binding site, which involves interactions with all four sugar rings in the glycolipid head group, contains a key tryptophan residue that packs against the B face of a galactose residue. The remainder of the site is generally polar. Many of the hydrogen bonds that form much of the protein–sugar interface involve basic and acidic amino acid side chains and the sugar hydroxyl groups. Differences in the structures of related forms of PapG explain the tropism of particular *E. coli* strains for different cell types (bladder or kidney) and different host species.

Protein–carbohydrate interactions play a part in many other bacterial adhesion systems, although none are understood at the same degree of molecular detail. The interaction of bacteria with host tissues can be complex. For example, the bacterium *Helicobacter pylori*, which is associated with the development of gastric ulcers, binds to at least two distinct oligosaccharide structures on different cell types in the stomach (Box 11.1).

 BOX 11.1   Glycobiology of disease: *Bacteria that cause stomach ulcers use blood group glycans as receptors*

The bacterium *Helicobacter pylori* colonizes the human stomach and is found in about half of the human population. In most people, infection with *H. pylori* does not produce clinical symptoms, but in some individuals, inflammation caused by the bacteria leads to development of stomach ulcers or stomach cancer. The factors that determine why some infected people develop disease are not yet known. *Helicobacter* binds to the mucous-secreting epithelial cells and surface mucins that line the stomach. This binding protects the bacteria from the acid environment of the stomach and helps to prevent them from being swept away into the stomach lumen.

The best-characterized mechanism of adhesion is through binding of proteins in the outer membrane of *Helicobacter* to fucosylated carbohydrate structures on the mucins. The *Helicobacter* adhesin BabA binds to the H-1 antigen that is found on glycans in secretions of individuals with blood group O, and to the related difucosylated Lewis[b] (Le[b]) antigen. People with blood group O are much more likely to develop stomach ulcers than individuals with other blood groups, suggesting that adhesion of *Helicobacter* to the H and Le[b] antigens may increase the severity of

Box 11.1   207

infection. Another *Helicobacter* protein, SabA, binds to sialyl-Lewis^x. Expression of sialyl-Lewis^x is not common on glycoproteins or glycolipids found in normal stomach tissue, but its expression is greatly increased during inflammation, including inflammation caused by *H. pylori* infection. It is thought that adherence of *Helicobacter* to sialyl Lewis^x contributes to the virulence and persistence of infection.

GlcNAcβ1-
3
Galβ1
2
Fucα1
H-1 antigen

Fucα1-4GlcNAcβ1-
3
Galβ1
2
Fucα1
Lewis^b antigen

Galβ1-4GlcNAcβ1-
3
Fucα1
Lewis^x antigen

Galβ1-4GlcNAcβ1-
2   3
Fucα1 Fucα1
Lewis^y antigen

Interestingly, *H. pylori* also expresses blood group glycans at its cell surface. Like other Gram-negative bacteria, *Helicobacter* has a cell wall containing lipopolysaccharides. Lipopolysaccharides are large molecules consisting of polysaccharide chains on an oligosaccharide core attached to lipid A. Lipid A is inserted into the outer membrane and the polysaccharides project from the cell surface. The structure of the carbohydrate part of lipopolysaccharide varies depending on the species and strain of bacteria. Most strains of *H. pylori* express structures identical to Lewis blood group antigens on their lipopolysaccharide with Lewis^x (Le^x) and Lewis^y (Le^y) antigens most commonly seen. The *H. pylori* genome encodes three fucosyltransferases needed for synthesis of the Lewis antigens.

Why *Helicobacter* synthesizes these carbohydrate structures is not yet clear. One possibility is that the Lewis antigens on lipopolysaccharide are involved in adherence of the bacteria through interactions with receptors on the stomach epithelial cells. However, recent experiments measuring binding of *H. pylori* to stomach epithelial cells indicate that lipopolysaccharide plays only a minor role in adhesion. Another possibility is that expression of carbohydrate structures identical to human antigens helps *H. pylori* to evade the host immune response, but again there is no clear evidence in favour of such a role. *H. pylori* strains expressing Le^x or Le^y are able to bind to the C-type lectin DC-SIGN (see Chapter 9) which is found at the surface of dendritic cells. *In vitro* experiments show that interaction of *H. pylori* with DC-SIGN reduces stimulation of T helper cells by dendritic cells. So it is possible that Lewis antigens on *Helicobacter* lipopolysaccharide help to suppress the immune response through binding to DC-SIGN on dendritic cells found in the stomach.

Stomach ulcers can usually be cured with antibiotics, but widespread use of antibiotics is not suitable for prevention or eradication of *Helicobacter* infection due to the risk of developing resistant strains. Other strategies are required, and possibilities include the development of vaccines against the *Helicobacter* carbohydrate-binding adhesins or the use of anti-adhesive carbohydrates.

Either approach might prevent colonization of the stomach by blocking the interactions between *Helicobacter* and the stomach mucous layer.

Essay topic

- Describe how glycan-binding adhesins of *Helicobacter* have been characterized. Assess the evidence that glycan–protein interactions are important in the development of stomach inflammation due to *Helicobacter* infection.

Lead references

- Bergmann, M.P., Engering, A., Smits, H.H., van Vliet, S.J., van Bodegraven, A.A., Wirth, H.-P., Kapsenberg, M.L., Vandenbroucke-Grauls, C.M.J.E., van Kooyk, Y., and Appelmelk, B.J. (2004). *Helicobacter pylori* modulates the T helper cell 1/T helper cell 2 balance through phase-variable interactions between lipopolysaccharide and DC-SIGN, *Journal of Experimental Medicine* **200**, 979–990.

- Karlsson, K.-A. (2000). The human gastric colonizer *Helicobacter pylori*: a challenge for host-parasite glycobiology, *Glycobiology* **10**, 761–771.

- Ilver, D., Arnqvist, A., Ögren, J., Frick, I.-M., Kersulyte, D., Incecik, E.T., Berg, D.E., Covacci, A., Engstrand, L., and Borén, T. (1998). *Helicobacter pylori* adhesin binding fucosylated histo-blood group antigens revealed by retagging, *Science* **279**, 373–377.

- Mahdavi, J., Sondén, B., Hurtig, M., Olfat, F.O., Forsberg, L., Roche, N. Ångström, J., Larsson, T., Teneberg, S., Karlsson, K.-A., Altraja, S., Wadström, T., Kersulyte, D., Berg, D.E., Dubois, A., Petersson, C., Magnusson, K.-E., Norberg, T., Lindh, F., Lundskog, B.B., Arnqvist. A., Hammarström, L., and Borén, T. (2002). *Helicobacter pylori* SabA adhesin in persistent infection and chronic inflammation. *Science* **297**, 573–578.

11.8   Viruses use lectins to target cell surfaces

Many viruses attach to cell surface glycans to initiate infection of eukaryotic cells. The best-understood examples involve binding to terminal sialic acid residues on cell surface glycoproteins and glycolipids. Interactions of surface proteins on the influenza virus with sialic acid-containing ligands have been the most extensively studied in molecular detail. The binding interaction that leads to infection by the influenza virus is mediated by a trimeric haemagglutinin protein that forms spikes on the outer surface of the lipid bilayer membrane encircling the virus capsid (Figure 11.10). The haemagglutinin gets its name from its ability to cross-link red blood cells in the same way that many plant lectins do. The haemagglutinin trimer is an elongated molecule with three sialic acid-binding sites at the ends of the molecule that point away from the viral membrane. In addition to sialic acid-binding activity, haemagglutinin has membrane fusion activity to initiate the process of entering the host cell.

The binding site in each haemagglutinin polypeptide is relatively shallow and interacts primarily with the terminal sialic acid residues linked to galactose.

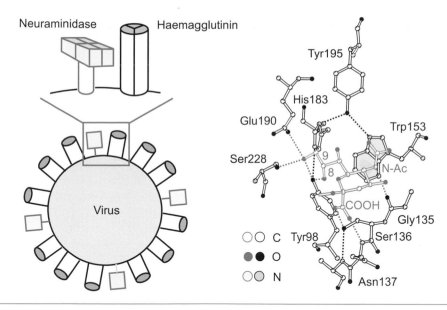

Figure 11.10 Influenza virus surface coat with trimeric haemagglutinin and tetrameric neuraminidase spikes. The sialic acid-binding site at the top of the haemagglutinin is shown in detail, highlighting hydrophobic packing of tryptophan against the methyl group on the 2-acetamido substituent and hydrogen bonds to the glycerol side chain and the carboxyl group. [Based on entry 1HGG in the Protein Databank.]

Although the overall fold of the haemagglutinin is not like any known bacterial, plant, or animal lectins, the principles that underlie cell surface recognition are similar to those already discussed. Key interactions in the binding site include the packing of the *N*-acetyl side chain against a tryptophan at the bottom of the binding site as well as polar and co-operative hydrogen bonds to the glycerol side chain and carboxyl group. The affinity for monomeric sialosides is weak (mM affinities), but binding to cell surfaces is enhanced by the simultaneous interaction of multiple haemagglutinin-binding sites with multiple sialic acid residues on the target.

A second molecule on the surface of influenza virus particles, the tetrameric neuraminidase (sialidase) molecule, forms smaller spikes on the surface. While the haemagglutinin is involved in entering a cell, the neuraminidase is required for the efficient release of newly made virus particles from an infected cell, since the virus particles would otherwise be retained by the haemagglutinin binding to sialic acid on the producing cell. It might seem to be counter-intuitive for the virus to carry an enzyme that destroys the receptors to which it binds. However, the large number of viruses released by an infected cell can efficiently remove the sialic acid and thus become free to move on to other cells, while the neuraminidase on a single virus arriving at a new host cell would be insufficient to remove all of the vast number of sialic acid residues on the target cell surface, ensuring that the haemagglutinin will be able to bind and initiate membrane fusion and entry. Influenza virus neuraminidase is the target for anti-viral drugs used to limit viral infections (Box 11.2).

BOX 11.2 Glycotherapeutics: *Anti-influenza drugs are neuraminidase inhibitors*

The active site of the neuraminidase is in a pocket that is substantially deeper than the sialic acid-binding site on haemagglutinin (Figure A). Correspondingly, the affinity of the neuraminidase for its substrate is in the micromolar range, while the affinity of the haemagglutinin for sialic acid is about a thousand-fold weaker, in the millimolar range. It is generally easier to design low molecular weight inhibitors that will make multiple favourable interactions in a deep pocket as in the neuraminidase, rather than a shallow groove as in the haemagglutinin. These factors make the neuraminidase a more attractive target than the haemagglutinin for design of compounds that can be used as drugs to prevent viral transmission, so the neuraminidase has been the primary target for anti-viral drug design.

Figure A

Haemagglutinin   Neuraminidase

In addition, because the neuraminidase is an enzyme, inhibitors designed to mimic the transition state in the hydrolysis of sialosides bind with particularly high affinity. During hydrolysis of the glycosidic linkage, the carbon at position 2 becomes planar, which can be mimicked by insertion of a double bond to this carbon (Figure B). Other changes made to create oseltamivir (Tamiflu), one of the currently approved anti-influenza drugs, were incorporated to make the product resistant to breakdown and to favour additional interactions with the active site of the neuraminidase, as well as to simplify the synthesis.

Figure B

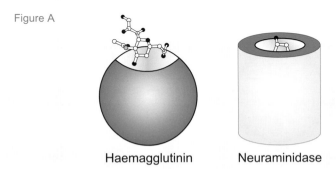

NeuAc   Transition state in hydrolysis reaction   Tamiflu

Neuraminidase inhibitors prevent efficient release of virus from an infected cell and hence inhibit spread of infection from cell to cell. Such drugs currently marketed for the treatment of influenza virus infections appear to provide some relief in the early stages of viral infection. There have been concerns about the ability of the virus to mutate under selective pressure from the inhibitors and thus become resistant to the action of the drugs. Many mutations that prevent drug binding would be expected to reduce the efficiency of the neuraminidase in releasing virus from infected cells. Although there is some evidence that resistant, but still virulent, forms of the virus can emerge, recent structural analysis of neuraminidases from multiple strains of virus may guide the design of a next generation of inhibitors.

Although it might seem that blocking entrance of the virus into the host would be most effectively achieved by blocking interaction of the viral haemagglutinin with host glycans, it would be difficult to make low molecular weight inhibitors of the haemagglutinin–receptor interaction. Only relatively large, oligomeric ligands that interact with multiple binding sites in the haemagglutinin will bind with sufficient affinity to block virus binding. Such large ligands are difficult to synthesize and deliver as orally active therapeutics.

Essay topic

- Discuss differences and common features in the design of oseltamivir (Tamiflu) and another commercially available anti-influenza drug, zanamivir (Relenza).

Lead references

- Gubareva, L.V. (2004). Molecular mechanisms of influenza virus resistance to neuraminidase inhibitors, *Virus Research* **103**, 199–203.

- Russell, R.J., Haire, L.F., Stevens, D.J., Collins, P.J., Lin, Y.P., Blackburn, G.M., Hay, A.J., Gamblin, S.J., and Skehel, J.J. (2006). The structure of H5N1 avian influenza neuraminidase suggests new opportunities for drug design, *Nature* **443**, 45–49.

- Wade, R.C. (1997). Flu' and structure-based drug design, *Structure* **5**, 1139–1145.

The primary factor in determining the cycles of epidemic and pandemic infections by influenza virus is evolution of antigenic sites on the surfaces of the haemagglutinin and neuraminidase. Such variations lead to the names of the different viral clades. For instance, the haemagglutinins of H1N1 and H5N1 strains have substantial antigenic differences and antibodies to one are not generally protective against the other. The mutations responsible for these differences generally lie far from the sialic acid binding site in the haemagglutinin and the active site of the neuraminidase.

However, another feature of the binding site of the haemagglutinin has direct relevance to the ability of viruses to move between species. Different strains of influenza virus bind preferentially to sialic acid in either α2-3 or α2-6 linkage, with the haemagglutinin specificity reflecting common forms of sialic acid on epithelial cells in the host. Human viruses bind efficiently to NeuAc in α2-6 linkage, which is found on epithelial cells in the trachea and other portions of the upper respiratory tract, while avian viruses bind better to NeuAc in α2-3 linkage as found in the intestine of birds. These differences in specificity form a barrier to the movement of viruses between species, although the fact that both α2-3 and α2-6-linked sialic acid are found in the trachea of pigs may explain why they can act as an intermediate host in which genes

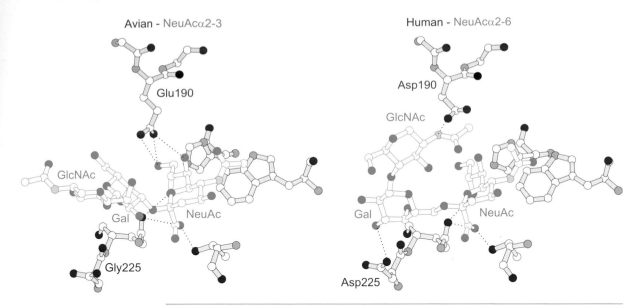

Figure 11.11 Structural differences between haemagglutinins of human and avian influenza viruses. The specificity of the human virus for NeuAcα2-6-Galα1-4GlcNAc is determined by favourable interactions of the *N*-acetyllactosamine part of the ligand with side chains Asp190 and Asp225, while the different orientation of the *N*-acetyllactosamine unit in NeuAcβ2-3-Galβ1-4GlcNAc means that the larger side chain of Glu190 is needed to interact favourably with the ligand and residue Gly225 no longer interacts with the ligand in the haemagglutinin from the avian virus. [Adapted from: Stevens, J., Blixt, O., Glaser, L., Taubenberger, J.K., Palese, P., Paulson, J.C., and Wilson, I.A. (2006). Glycan microarray analysis of the haemagglutinins from modern and pandemic influenza viruses reveals different receptor specificities, *Journal of Molecular Biology* **355**, 1143–1155 and based on entries 1RVX and 1RVZ in the Protein Databank.]

from human and avian species mix. Just two changes in amino acid sequence between the binding sites of the avian-type and human-type neuraminidases determine the difference in specificity for α2-3 or α2-6-linked sialic acid (Figure 11.11), suggesting that viruses may be able to alter their host range with only very limited genetic change.

11.9   Lectins appeared early in evolution but have diverse functions in higher organisms

The fact that lectins are found in bacteria and plants as well as animals suggests that they appeared early in evolution. The genome sequences for vertebrates, invertebrates, plants, and both eukaryotic and prokaryotic micro-organisms can be compared to determine when different structural classes of lectins evolved (Figure 11.12). Calnexin and calreticulin are found in all multicellular eukaryotic organisms and in some types of yeasts as well. The role of these proteins in glycoprotein quality control seems to have evolved at about the time the N-linked glycosylation pathway

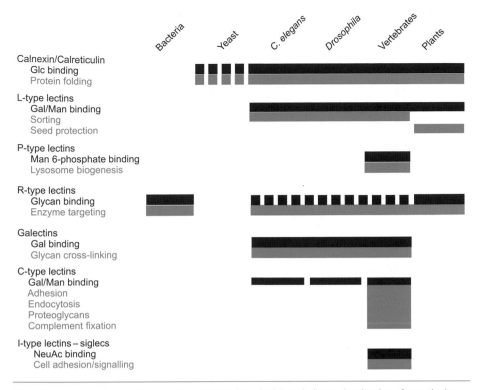

Figure 11.12 Evolution of different groups of lectins. Black bars indicate that lectins of a particular type with sugar-binding activity are present. Blue bars indicate that biological functions for these lectins are known.

took shape. Thus, providing a tag for sorting and quality control was probably one of the earliest functions for this pathway in eukaryotic cells.

The distribution and functions of other luminal lectins in multicellular eukaryotes vary widely. For example, L-type lectins in animals are involved in protein sorting, while in plants they play a part in host defence. R-type CRDs are the only sugar-binding domains that have been identified in both prokaryotes and eukaryotes. The high degree of conservation in structure and in sugar-binding function suggests that a gene encoding an R-type CRD may have recently moved laterally between species. Domains related in structure to the mannose 6-phosphate receptors are found in invertebrate and yeast proteins, but these domains do not have the residues that form the sugar-binding site. The fact that lysosomes can be generated in the absence of these receptors suggests that there must be alternative routing signals that have been superseded in vertebrate cells.

Two major groups of extracellular animal lectins, the galectins and the C-type lectins, are found in both invertebrates and vertebrates but not in plants or yeasts. These two groups of animal-specific lectins show distinct patterns of evolution (Figure 11.13). Like many of the luminal lectins, the galectins are relatively conserved in all species where they have been examined. In contrast, the C-type lectins radiated

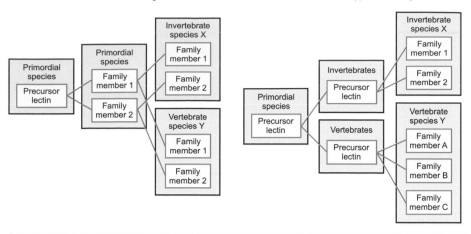

Figure 11.13 Alternative evolutionary paths of different types of lectins. For many luminal lectins and galectins, it is possible to identify equivalent proteins (orthologues) in vertebrates and invertebrates. In contrast, none of the vertebrate C-type lectins is a specific orthologue of any one of the invertebrate C-type lectins.

independently in vertebrates and invertebrates. Although C-type CRDs are a common feature of animal genomes, they probably serve different functions in different animals. The siglecs are found exclusively in vertebrates. Thus, some recognition functions of sugars are ancient and universal components of eukaryotes, while others are recent elaborations unique to specific groups such as the vertebrates.

Despite the fact that CRDs in different families of lectins have different overall folds, some shared principles of carbohydrate recognition are evident. Modest affinities for monosaccharide ligands are achieved with relatively few interactions, which include hydrophobic packing as well as formation of hydrogen and coordination bonds with characteristic hydroxyl groups in different sugar ligands. Specificity is achieved by a combination of accommodation of one type of ligand and exclusion of other types. Higher affinity and selectivity is achieved in individual CRDs through extended binding sites and in lectin oligomers by clustering of multiple binding sites. In such cases, increased valency leads to higher affinity, while the geometry of the oligomer provides selectivity for specific types of ligands. These features of sugar-binding sites mean that they can appear in a variety of protein frameworks and are difficult to detect from simple sequence analysis. Thus, many additional families of lectins may well remain to be described.

SUMMARY

The use of oligosaccharides derived from cell walls as signalling molecules in plants and the adoption of mammalian cell surface glycans as receptors for bacteria and viruses illustrate how glycans have been adapted to serve new functions during evolution. It is interesting to note that many of the non-mammalian lectins described in this chapter bind to mammalian glycan structures similar or identical to structures recognized by mammalian lectins. The L-type CRDs in the legume lectins and the R-type CRDs in ricin provide evidence for a common origin for some mammalian and non-mammalian sugar-binding proteins. In other cases, sugar-binding activity seems to have arisen independently in the context of novel protein folds. Nevertheless, the themes of shallow, weak monosaccharide-binding sites that are extended and combined to achieve higher affinity is repeated in each branch of organisms and the sugar–protein interactions rely on many of the same types of hydrogen bonding and hydrophobic interactions. From a biological point of view, it is important to remember that mammalian glycans have co-evolved not just with endogenous receptors but also with the various types of non-mammalian glycans and receptors discussed in this chapter. In some cases, selection might favour changes to help evade the effects of pathogens but in other instances conservation of glycans would be required to maintain mutually beneficial interactions.

KEY REFERENCES

Beddoe, T., Paton, A.W., Le Nours, J., Rossjohn, J., and Paton, J.C. (2010). Structure, biological functions and applications of the AB$_5$ toxins, *Trends in Biochemical Sciences* **35**, 411–418. An overview of carbohydrate-binding bacterial toxins, including details of crystal structures, is presented.

Bouckaert, J., Hamelryck, T., Wyns, L., and Loris, R. (1999). Novel structures of plant lectins and their complexes with carbohydrates, *Current Opinion in Structural Biology* **9**, 572–577. An overview of the structural basis for carbohydrate recognition by different types of plant lectin is provided.

Day, P.J., Owens, S.R., Wesche, J., Olsnes, S., Roberts, L.M., and Lord, J.M. (2001). An interaction between ricin and calreticulin that may have implications for toxin trafficking, *Journal of Biological Chemistry* **276**, 7202–7208. Experiments demonstrating that calreticulin acts as a carrier for ricin during retrograde transport to the endoplasmic reticulum are described.

Dodd, R.B. and Drickamer, K. (2000). Lectin-like proteins in model organisms: implications for evolution of carbohydrate-binding activity, *Glycobiology* **11**, 71R–79R. A comparative analysis of lectins identified in the genomes of yeast, *Drosophila*, and *Caenorhabditis elegans* is presented.

Dodson, K.W., Pinkner, J.S., Rose, T., Magnusson, G., Hultgren, S.J., and Waksman, G. (2001). Structural basis of the interaction of the pyelonephritic *E. coli* adhesin to its human kidney receptor, *Cell* **105**, 733–743. Analysis of the crystal structure of PapG in complex with its globoside ligand, providing insights into this host–pathogen interaction is described.

Hamel, L.-P. and Beaudoin, N. (2010). Chitooligosaccharide sensing and downstream signalling: contrasted outcomes in pathogenic and beneficial plant-microbe interactions, *Planta* **232**, 787–806. An up-to-date review of the molecular events involved in oligosaccharide signalling in plants is presented.

Linton, D., Dorrell, N., Hitchen, P.G., Amber, S., Karlyshev, A.V., Morris, H.R., Dell, A., Wren, A.B., and Aebi, M. (2005). Functional analysis of the *Campylobacter jejuni* N-linked protein glycosylation pathway', *Molecular Microbiology* **55**, 1695–1703. Experiments defining the glycans and enzymes in the N-linked glycosylation pathway of Campylobacter are presented.

Lis, H. and Sharon, N. (1998). Lectins: carbohydrate-specific proteins that mediate cellular recognition, *Chemical Reviews* **98**, 637–674. An extensive review of lectins including a wealth of detail on lectins from plants and micro-organisms is presented.

Lord, J.M. and Roberts, L.M. (1998). Retrograde transport: going against the flow, *Current Biology* **8**, R56–R58. This is a short commentary on transport from the Golgi back to the endoplasmic reticulum.

Promé, J.-C. (1996). Signalling events elicited in plants by defined oligosaccharide structures, *Current Opinion in Structural Biology* **6**, 671–678. A detailed review of oligosaccharides as signalling molecules in plants is given.

Sandvig, K. and van Deurs, B. (2000). Entry of ricin and Shiga toxin into cells: molecular mechanisms and medical perspectives, *EMBO Journal* **19**, 5943–5950. A detailed review of the mechanism of action of these two carbohydrate-binding toxins is presented.

Sauer, F.G., Barnhart, M., Choudhury, D., Knight, S.D., Waksman, G., and Hultgren, S.J. (2000). Chaperone-assisted pilus assembly and bacterial attachment, *Current Opinion in Structural Biology* **10**, 548–556. The structure and function of adhesins of bacterial pili are reviewed in detail.

Steeves, R.M., Denton, M.E., Barnard, F.C., Henry, A., and Lambert, J.M. (1999). Identification of three oligosaccharide binding sites in ricin, *Biochemistry* **38**, 11677–11685. Affinity labelling studies of ricin are presented.

Stevens, J., Blixt, O., Glaser, L., Taubenberger, J.K., Palese, P., Paulson, J.C., and Wilson, I.A. (2006). Glycan microarray analysis of the hemagglutinins from modern and pandemic influenza viruses reveals different receptor specificities, *Journal of Molecular Biology* **355**, 1143–1155. Analysis of the structural basis for recognition of sialic acid in specific linkages by influenza haemagglutinin is described.

QUESTIONS

11.1   Discuss the different ways that glycans are involved in diseases caused by bacteria, viruses, and parasites.

11.2   How does N-linked glycosylation in *Campylobacter* compare to N-linked glycosylation in eukaryotes? What are the prospects for adapting this system for production of therapeutic glycoproteins in bacteria?

References: Schwarz, F., Hunag, W., Li, C., Schulz, B.L., Lizak, C., Palumbo, A., Numao, S., Neri, D., Aebi, M., and Wang, L.-Xi (2010). A combined method for producing homogeneous glycoproteins with eukaryotic N-glycosylation, *Nature Chemical Biology* **6**,264–266. Szymanski, C.M., Logan, S.M., Linton, D., and Wren, B.W. (2003). *Campylobacter* – a tale of two protein glycosylation systems, *Trends in Microbiology* **11**, 233–238.

11.3   Discuss how studies of the interactions of glycan-binding receptors on pathogens provide targets for design of antimicrobial drugs.

Lead reference: Von Itzstein, M. (2008). Disease-associated carbohydrate-recognizing proteins and structure-based inhibitor design, *Current Opinion in Structural Biology* **18**, 558–566.

11.4   Transgenic plants are attractive as hosts for producing recombinant glycoproteins. Discuss why the glycosylation produced by plant cells might not be suitable for therapeutic glycoproteins and what can be done to overcome this problem.

Lead reference: Strasser, R., Altmann, F., Mach, L., Glossl, J., and Steinkellner, H. (2004). Generation of Arabidopsis thaliana plants with complex N-glycans lacking beta1,2-linked xylose and core alpha1,3-linked fucose, *FEBS Letters* **561**, 132–136.

11.5   Compare the molecular mechanisms by which different carbohydrate-recognition proteins recognize sugars. Include in your discussion proteins from animals, plants, bacteria, and viruses.

12 Glycobiology and development

LEARNING OBJECTIVES

By the end of this chapter, students should understand:

1   The mechanisms of proteoglycan modulation of growth factor activities

2   Genetic evidence for the roles of proteoglycans in the development of vertebrates and invertebrates

3   How O-fucose glycans regulate patterning early in development

4   Multiple ways in which glycolipids may control development of the nervous system

5   How changes in glycosylation of T cells regulate positive and negative selection

Much of our knowledge of the biological functions of glycosylation derives from observing the phenotypes that result from mutations in genes encoding glycosyltransferases and glycan-binding receptors. The fact that many of these mutations result in developmental abnormalities provides compelling evidence for the importance of glycosylation in many stages in the development of multicellular organisms. In a few cases, we have a relatively complete understanding of how particular genetic changes affect development and several of these examples are discussed in this chapter. The broader implications of the effects of mutations that are not fully understood are also discussed. Unfortunately, in these instances our knowledge is more descriptive than mechanistic. In addition to the roles of glycans in early embryonic development and formation of the body plan of animals, functions in neural development and the immune system are considered.

12.1   Biochemical analysis has demonstrated how cell surface proteoglycans serve as co-receptors for growth factors

Biochemical analysis in mammalian systems combined with genetic analysis in *Drosophila* has established multiple important functions for proteoglycans in development, particularly those containing heparan sulphate chains. These roles include:

- functioning as co-receptors for stimulation of receptor kinases;
- sequestration of growth factors to prevent receptor stimulation;
- transport and stabilization of gradients for growth factors and morphogens.

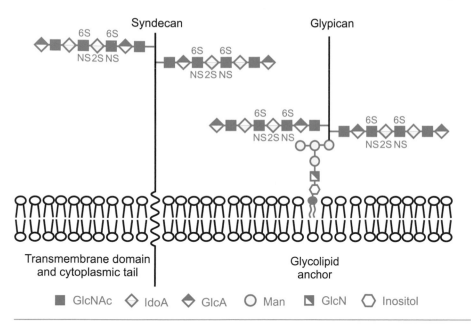

Syndecan   Glypican

Transmembrane domain
and cytoplasmic tail

Glycolipid
anchor

■ GlcNAc   ◇ IdoA   ◆ GlcA   ○ Man   ◨ GlcN   ⬡ Inositol

Figure 12.1 Organization of membrane proteoglycans. Syndecans are transmembrane proteins that contain several glycosaminoglycan attachment sites. Glypicans are attached to the plasma membrane through C-terminal glycolipid anchors and bear glycosaminoglycans attached within 50 amino acids of the C-terminus.

Some of these functions are mediated by matrix proteoglycans, while others involve a second major class of proteoglycans that reside in plasma membranes (Figure 12.1). The extracellular domains of these core proteins bear one or a few glycosaminoglycan chains. They can be anchored to the membrane by a transmembrane domain, in the case of syndecans, or by a glycolipid anchor, in the case of glypicans. These core proteins are always conjugated to glycosaminoglycan chains, but others are part-time proteoglycans and exist in two forms, with and without glycosaminoglycans attached.

The heparan chains attached to syndecans are essential for signalling by fibroblast growth factor (FGF) (Figure 12.2). The proteoglycan serves as an essential co-receptor for the growth factor to allow activation of the primary receptor. The extracellular domain of the primary receptor contains binding sites for the growth factor in terminal immunoglobulin-type modules, while the intracellular domain has tyrosine kinase activity. Analysis of co-crystals of growth factor with heparan sulphate and the receptor suggests mechanisms by which heparan sulphate may stabilize the ternary complex. In one model, two growth factor molecules, each interacting with a receptor subunit, are held together by binding to heparan sulphate. These interactions, plus interaction of heparan sulphate with one subunit of the receptor, stabilize the dimeric form of the receptor and lead to intracellular signalling events.

The segment of a heparan sulphate chain that interacts with a growth factor and its receptor must meet specific structural requirements. There are often clusters of

➲ See section 3.5 for more on matrix proteoglycans.

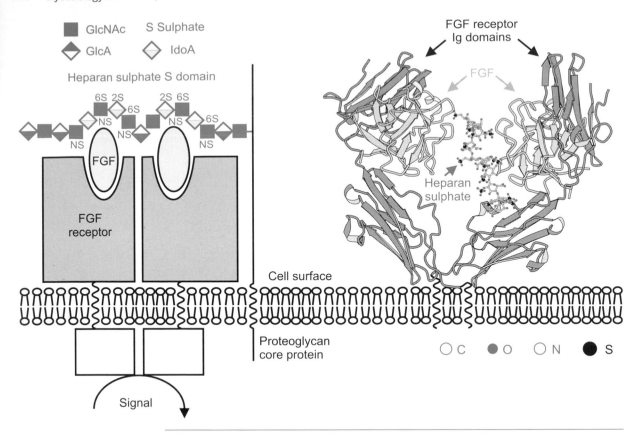

Figure 12.2 Function of cell surface heparan sulphate proteoglycan as a co-receptor for fibroblast growth factor (FGF). Signalling requires dimerization of the receptor–FGF complex. The dimer complex is stabilized by specific sequences of sulphated residues in heparan sulphate. One complex seen by X-ray crystallography is shown. [Based on entry 1E00 in the Protein Databank.] Ig, immunoglobulin.

⊙ See section 3.6 for more on heparan sulphate synthesis.

⊙ See Box 6.1 for more on heparin.

modified regions within proteoglycan chains. In heparan sulphate, such regions contain iduronic acid residues derived from glucuronic acid by epimerization, as well as N- and O-linked sulphate residues. Different patterns of heparan sulphate modification in different tissues and at different times of development provide a possible mechanism for regulation of growth factor activity. However, in contrast to the strict requirements for a specific sequence of sugar modification in heparin in order for it to activate antithrombin, the ability of heparan sulphate to act as a co-receptor for many members of the fibroblast growth factor family is largely correlated with overall charge density rather than any single modification.

Different patterns of modifications are required for other interactions of glycosaminoglycan chains with proteins. For example, the protein fibronectin, which mediates interactions between cells and the surrounding matrix, binds both to integrin and proteoglycan receptors on cell surfaces. The interaction with specially modified portions of the glycosaminoglycan chains of the proteoglycan again occurs in a positively charged groove on the protein surface, suggesting that stabilization of protein–protein interactions by binding to appropriately aligned regions of positive charge on two proteins may be a common mechanism for proteoglycan function.

12.2   Mutant mice provide evidence for the roles of proteoglycans in mammalian development

One approach to genetic analysis of development is to undertake random muta-genesis and select mutants that fail to develop normally. In mice, the *lazy mesoderm* mutation, which was identified in such an unbiased screen, is in a gene required for proteoglycan synthesis, providing important evidence that proteoglycans are critical for development. The mutant gene encodes UDP-glucose dehydrogenase, which is needed to make the UDP-glucuronic acid used in the synthesis of glycosaminogly-cans (Figure 12.3). As suggested by its name, the *lazy mesoderm* mutation causes failure of mesoderm migration during gastrulation and is therefore embryonic lethal. This phenotype is similar to that observed in mice lacking fibroblast growth factor 8, and further analysis confirms that the defect in cell migration is due to a failure of fibroblast growth factor signalling. This mutation thus provides *in vivo*

⊙ See **section 5.8** for more on knockout mice and development.

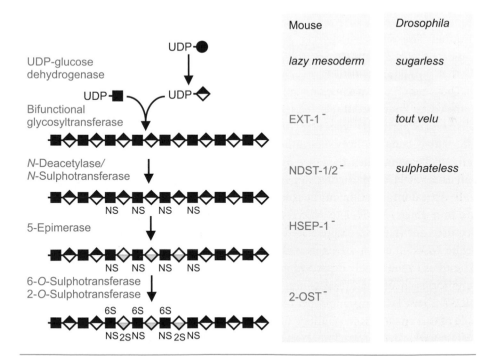

Figure 12.3 Mutations in the pathways for biosynthesis of proteoglycans. The enzymes defective in mouse and *Drosophila* mutations which cause developmental defects are indicated. Mutations that were obtained from random mutagenesis are indicated in black and those obtained by gene knockout techniques are indicated in colour.

evidence that proteoglycans function as cofactors for growth factor signalling which complements the biochemical and cell biological studies *in vitro*.

Creation of mice in which selected genes have been disrupted provides another powerful way to test the functions of glycans. Further evidence for the importance of proteoglycans has been obtained by knockout of the gene encoding the bifunctional glycosyltransferase required for synthesis of the repeating disaccharide of heparan sulphate. Mice lacking this gene fail to undergo gastrulation, again reflecting an early role for heparan sulphate in tissue migration in the embryo. Genes required for generation of the specifically modified clusters of sugar residues believed to be required for binding to fibroblast growth factors and their receptors have also been targeted. The glucuronyl C5 epimerase and the 2-O-sulphotransferases are each encoded by single genes, so single knockouts have been used to prevent these modifications. Both mutations are lethal, as a result of significant developmental abnormalities, but the mice develop much further than those completely unable to synthesize glycosaminoglycans. Knockout of the 2-O-sulphotransferases results in a lack of kidneys and malformed eyes and skeleton, while the epimerase is essential for proper kidney, lung, and skeletal development. However, other organ systems develop normally in spite of evidence that multiple heparan sulphate-dependent growth factors are required for earlier stages in patterning of the embryo. Compensating increases in 2-N and 6-O sulphation generate heparan sulphate that is able to interact with growth factors, consistent with the biochemical evidence that many of these interactions are more related to overall negative charge than to a particular pattern of modification. The restricted defects seen in the mice with altered heparan sulphate indicate that there are specific roles for the various forms of heparan sulphate, but further genetic and biochemical analysis is required to define these roles.

The phenotypes of knockout mice must always be interpreted with caution. In some cases, knockout of selected transferases has not resulted in the expected complete loss of particular types of glycans, providing an important route to discovery of alternative, compensatory biosynthetic pathways. Knockout studies can be particularly challenging when there are multiple genes encoding a family of enzyme that can each catalyse a particular step in glycan biosynthesis, but mutant mice can provide critical information on the roles of the individual enzymes. For example, there are four genes, *NDST-1* to *NDST-4*, encoding *N*-deacetylase/*N*-sulphotransferases required for heparan sulphate biosynthesis. NDST-1 and 2 are widely expressed while NDST-3 and 4 are expressed only in restricted tissues during development. The roles of the widely expressed enzymes have been examined by making knockouts for each of these two genes. The most obvious effect of knocking out the *NDST-2* gene is to prevent proper mast cell development, indicating that the primary role of this enzyme is in synthesis of the highly sulphated heparin found in the mast cell granules. In contrast, knockout of NDST-1 results in substantial reduction in the amount of heparan sulphate in embryos and causes death in late embryogenesis or just after birth. Again, in spite of major alteration to the heparan sulphate complement in these mice, basic embryonic patterning and organ formation takes place. However, double knockout mice lacking both NDST-1 and 2 resemble those completely unable to make the heparan polymer at all, because they fail to undergo proper gastrulation. Thus, although NDST-1 and NDST-2 normally modify heparan at different sites and in distinct stages of development, the presence of either one of the enzymes provides sufficient modified heparan to allow development up to birth.

12.3   Study of *Drosophila* and other model organisms reveals multiple roles for heparan sulphate proteoglycans

Our understanding of many aspects of mammalian biochemistry has benefited from the study of simpler model organisms. Analysis of glycans and glycan-binding proteins in such model organisms suggests that some of the functions of glycans are amenable to study in this way. Invertebrates and simpler vertebrates, such as *Drosophila*, *Caenorhabditis elegans*, *Xenopus* and zebrafish, have been extensively investigated by developmental biologists and they provide examples of the types of functions that glycans can perform. Some roles of glycans are directly comparable in model organisms and mammals, and in such cases, the genetic tractability of the simpler organisms makes them very powerful systems in which to study the functions of glycans.

Screens of *Drosophila* for mutants that show abnormal development have led to the identification of mutations in several genes involved in proteoglycan biosynthesis, indicating that proteoglycans play a critical role in development in invertebrates as well as vertebrates. Genes identified in this way include those that encode proteoglycan core proteins and enzymes required for heparan sulphate synthesis (Figure 12.3). For example, mutations in either the gene *dally-like*, which encodes one of the *Drosophila* glypicans, or the gene for syndecan result in failure of correct axon guidance in embryogenesis and mutations in *dally*, the gene for the other *Drosophila* glypican, disrupt morphogenesis in multiple tissues. Genetic studies indicate that these effects are due to failure of the mutant forms of Dally to interact with growth factors Wingless, a member of the Wnt family, and Decapentaplegic, a transforming growth factor β/bone morphogenetic protein family member. Although the nature of the interactions of these growth factors are less well understood than for fibroblast growth factor, both are heparan sulphate binding proteins.

Direct evidence for the importance of the heparan sulphate portion of the proteoglycans has been obtained from disruption of wingless signalling by mutations in the gene *sugarless*, which is analogous to the mouse *lazy mesoderm* mutation and results in the inability to synthesize UDP-glucuronic acid and thus compromises the ability to make any of the glycosaminoglycan chains. Disruption of Wingless signalling also results from reduction in heparan sulphate synthesis caused by mutation of *tout velu*, which encodes the bifunctional glycosyltransferase that makes the backbone polymer, and in *sulphateless*, which encodes the only heparan *N*-deacetylase/ *N*-sulphotransferases in *Drosophila*.

The effects of some of the mutations in proteoglycan biosynthesis may result from the need for properly presented heparan sulphate chains to function as co-receptors for growth factors in the fibroblast growth factor, Wnt and bone morphogenetic protein families, but some of the phenotypes reflect other roles of the proteoglycans that can be demonstrated by examining the effects of sulphatases on the surfaces of cells. These enzymes can remove some of the sulphate from heparan sulphate and modulate its activity. For example, the enzymes Sulf1 and Sulf2 selectively cleave 6-O-sulphate linkages. In *Xenopus*, action of Sulf1 on cell surface heparan sulphate inhibits the co-receptor function of the endogenous heparan sulphate, probably because the 6-O-sulphate groups are required for the interaction with the fibroblast growth factor receptor, although binding to fibroblast growth factor itself is not affected. However, the same enzyme stimulates interaction of Wnt proteins with their frizzled receptors. It has been suggested that the removal of sulphate residues reduces the affinity of heparan sulphate for the Wnt protein, allowing it to move to the receptor and activate it (Figure 12.4).

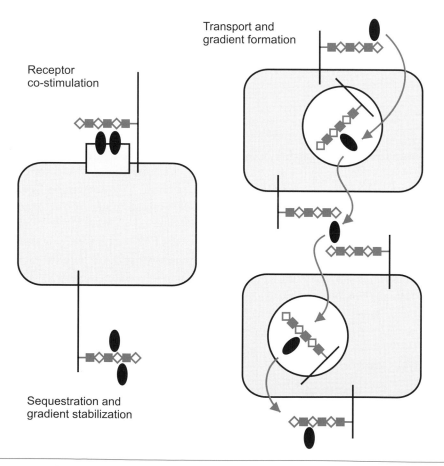

Figure 12.4 Alternative mechanisms of proteoglycan-growth factor interactions. Growth factors are depicted as black ellipses and glycosaminoglycans are shown highlighted in colour.

Many growth factors are basic and thus adhere to the glycosaminoglycan chains of both cell surface and extracellular matrix proteoglycans, leading to the type of sequestration observed for the Wnt proteins. By preventing diffusion, interactions with proteoglycans thus stabilize concentration gradients of growth factors and can also increase their local concentrations. Sulphatases may provide a mechanism for releasing them and initiating activation of membrane receptors. This arrangement can be important, because unlike hormones that travel throughout the body, growth factors usually function over relatively short distances. Sulf1 and Sulf2 can also potentiate Wnt signalling in cancer cells, probably by a similar mechanism of mobilization, and may be required for release of growth factors to stimulate cell growth at sites of wound repair.

The phenotype of mutations in the *tout velu* gene of *Drosophila* suggests still another role for proteoglycans in the control of growth factor distribution. These mutations disrupt pattern formation controlled by the hedgehog protein. Although hedgehog is secreted and cells can respond to it, the growth factor fails to migrate away from the cell in which it is synthesized and hence it stimulates immediately neighbouring cells, but does not reach other cells within a radius of up to ten cells as it does in the wild-type fly. It is believed that hedgehog protein binds to heparan sulphate on proteoglycans and may be transported through cells to reach their neighbours. Thus, interactions with proteoglycans can enhance as well as limit mobility of growth factors.

Although there are examples of conserved glycan functions, many invertebrate and vertebrate glycans and glycan-binding receptors differ substantially. Glycans that are found in only some groups of organisms have probably evolved to perform special functions. For example, core α1-3-linked fucose residues of the type found in invertebrates but not in vertebrates are detected in the developing brain of *Drosophila* along with a receptor specific for this structure, leading to the suggestion that interactions between fucosylated glycans and the receptor help to organize the developing nervous system. Because such fucosylated glycans are absent from mammals, the *Drosophila* receptor is not a good model for a specific receptor in mammals.

 BOX 12.1 Glycobiology of disease: *Human diseases result from aberrant proteoglycan biosynthesis*

Mutations in human homologues of several of the mouse and *Drosophila* genes discussed in this chapter have been identified in human patients with developmental abnormalities. Although proteoglycan genes are linked to developmental defects in a number of species, it is interesting that the phenotypes associated with mutations in homologous genes are often quite different. Investigation of glycosylation changes associated with human developmental disorders provides an important source of information about glycan function that is often complementary to studies of mice and other model organisms.

Diastrophic dysplasia, literally twisted misgrowth, is a form of dwarfism in which cartilage biosynthesis is defective. Undersulphation of matrix proteoglycans results from defective transport of sulphate. The most severe form of the disease is caused by the complete absence of the diastrophic dysplasia sulphate transporter encoded by the *DTDST* gene, while missence mutations that result in intermediate levels of transport are associated with milder symptoms. It is possible that the reduction in sulphation of heparan sulphate on cell surface proteoglycans leads to reduced activity of growth factors, as do mutations in the *sulphateless* gene in *Drosophila*. However, the formation of defective cartilage is usually attributed to loss of sulphation of matrix proteoglycans such as aggrecan. Even homozygous null mutations in the *DTDST* gene do not result in complete loss of sulphate transport in all cell types, so it is possible that cell surface proteoglycans are less affected by the reduced transport.

The disease hereditary multiple exostoses results in development of benign tumours on the growth plates of bone. Two of the *EXT* genes linked to this disease encode bifunctional polymerases that synthesize the disaccharide repeats of heparan sulphate and the disease-causing mutations result in loss of polymerase function. By analogy to the *tout velu* mutants of *Drosophila*, it might be expected that tumour development results from disruption of growth factor activity because of decreased levels of heparan sulphate on cell surface proteoglycans. However, it is not possible to draw a simple analogy because the disease phenotype suggests increased rather than decreased growth factor activity in the absence of appropriate heparan sulphate synthesis.

There are six distinct glypicans in humans, compared to the two in *Drosophila*. Mutation of the human *glypican-3* gene causes Simpson–Golabi–Behmel syndrome, characterized by overgrowth of tissues and susceptibility to tumour formation. Based on the analogy with mutations in the *Drosophila* glypican genes, it is tempting to speculate that the human syndrome reflects a role of glypican-3 in action of a growth factor, but the molecular basis for the phenotype remains to be explained. As in the case of the *EXT* genes, the human mutations appear to increase growth factor activity rather than decrease it.

Essay topics

- Compare the effects of mutations in *Drosophila*, mouse, and human genes that encode enzymes that synthesize the core glycosaminoglycan chain of heparan sulphate.
- Discuss the proposed roles for glypicans in activation of Wnt family members.

Lead references

- Karniski, L.P. (2004). Functional expression and cellular distribution of diastrophic dysplasia sulphate transporter (DTDST) gene mutations in HEK cells, *Human Molecular Genetics* **13**, 2165–2171.
- Nadanaka, S. and Kitagawa, H. (2008). Heparan sulphate biosynthesis and disease, *Journal of Biochemistry (Tokyo)* **144**, 7–14.
- Song, H.H., Shi, W., Xiang, Y.-Y., and Filmus, J. (2005). The loss of glypican-3 induces alterations in Wnt signaling, *Journal of Biological Chemistry* **280**, 2116–2125.

12.4 O-linked fucose-based glycans are important for extracellular signalling during development in vertebrates and invertebrates

Fucose attached directly to serine or threonine plays an important part in developmental tissue patterning in both vertebrates and invertebrates. The role of this modification has been studied most extensively in the cell surface Notch receptor in *Drosophila*. Notch comprises two subunits, an extracellular subunit and a transmembrane and intracellular subunit. When stimulated by interactions with an appropriate ligand on a nearby cell surface, the cytoplasmic domain of Notch is cleaved, releasing a peptide that migrates to the nucleus and initiates a programme of gene transcription.

Notch-controlled gene expression occurs in cells located at the midline in developing wings. Although Notch is expressed throughout the wing imaginal disk,

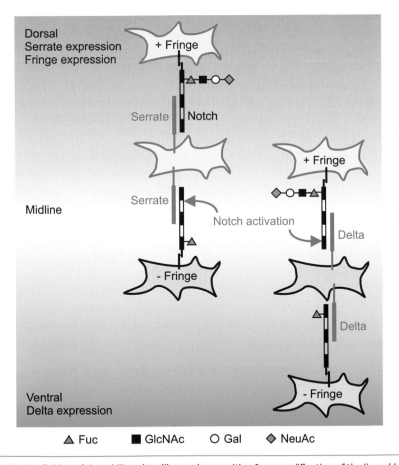

Figure 12.5 Definition of the midline signalling region resulting from modification of the ligand-binding activity of Notch by the Fringe glycosyltransferase. Signalling from the Notch protein in response to different ligands depends on its state of glycosylation.

which is the group of cells that will form the wing, clear demarcation of the midline cells is achieved by selective activation of Notch at the interface between the dorsal and ventral regions. Activation results from interaction with different receptors in the regions on either side of the midline (Figure 12.5). Two different ligands can activate Notch. The ligand Delta is expressed in the ventral region while an alternative ligand, Serrate, is expressed in the dorsal region, along with the product of the *fringe* gene. Fringe differentially modulates the ability of Notch to interact with these two ligands. In the ventral region, Notch expressed alone is unresponsive to Delta and in the dorsal region, Notch co-expressed with Fringe is unresponsive to Serrate. However, at the midline, cells from the dorsal side expressing Notch and Fringe encounter Delta on cells from the ventral side, leading to Notch activation. Conversely, ventral cells expressing Notch only encounter dorsal cells expressing Serrate, also leading to Notch activation.

The change in the response of Notch to the Delta and Serrate ligands is due to the ability of Fringe to modify O-linked glycans on Notch. The extracellular portion of Notch consists of 36 epidermal growth factor-like modules (Figure 12.6). In many of these modules, a sequence motif directs attachment of fucose to a specific threonine or serine residue. The presence of these fucose residues is required for Notch to reach the cell surface. The *fringe* gene encodes a GlcNAc transferase that initiates further elaboration of these residues with GlcNAc, galactose, and sialic acid residues, the presence of which change the response to interaction with Delta and Serrate. The presence of additional unusual O-linked glucose residues is also required for Notch to function. The molecular mechanism by which glycans modify the interaction of Notch with Delta and Serrate is not yet understood, but at least one of the Fringe-dependent glycans is attached to epidermal growth factor domain 12, which lies in the portion of Notch that interacts with these ligands (Figure 12.6). This change represents one of the most specific effects on protein function that has been associated with a change in glycan structure. Analogous modifications to vertebrate homologues of Notch, which are also involved in pattern formation in early embryos, suggest that the roles of these glycans are preserved throughout the animal kingdom.

12.5   Cell surface glycolipids are important for the development of the nervous system

Glycolipids form less than 5% of the lipid content of most tissues, but represent more than 25% of the lipid content of the myelin sheaths that insulate axons in the nervous system. The myelin sheaths are formed from the plasma membranes of Schwann cells and oligodendrocytes that wrap repeatedly around the axons. The layers of membrane undergo compaction, in which the cytoplasmic surfaces pack closely against each other (Figure 12.7). Interaction of the extracellular surfaces of the membranes completes compaction. Galactose-based galactosylceramide and sulphatide are particularly abundant in the myelin sheaths, but an enormous variety of glucosphingolipids are also present. The glycan head groups of the glycolipids could

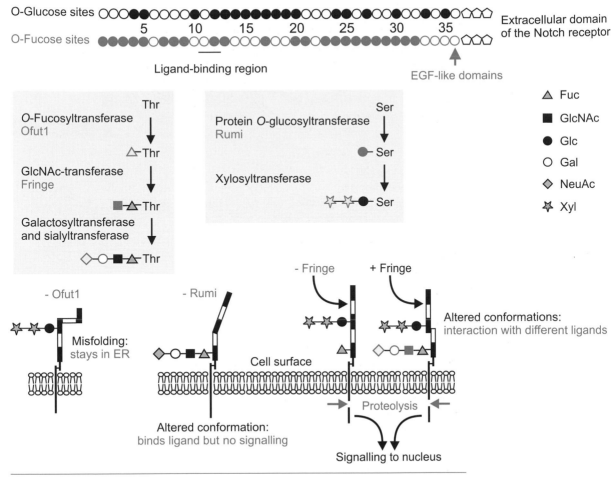

Figure 12.6 Modification of Notch behaviour by glycosyltransferases. Fucose residues added to epidermal growth factor (EGF) domains of Notch by O-Fucosyltransferase 1 are required for the Notch protein to fold correctly and exit the endoplasmic reticulum (ER). Glucosylation by Rumi followed by addition of xylose residues is also required for Notch to assume a conformation able to mediate intracellular signalling. The fucose residues can be elongated by Fringe, which adds GlcNAc residues that are then further extended. The presence of this elongated glycan alters the interactions of Notch with different ligands.

mediate several aspects of myelination, including interactions between the myelin layers, interactions between the innermost layer of myelin, and the axon or interactions at the edges of the myelin at the gaps known as the **nodes of Ranvier**.

Several types of knockout mice have been created to probe the roles of glycolipids in the nervous system. Individual cells lacking the enzyme that mediates glucosylceramide synthesis multiply normally, indicating that these glycolipids are not essential to basic cell physiology. However, mouse embryos lacking this transferase fail to develop beyond the eight-day stage, reflecting the importance of at least some of the glucosphingolipids in development. Mice that can make glucosylceramide but lack the GalNAc-transferase that extends the ganglio series of glycolipids are viable.

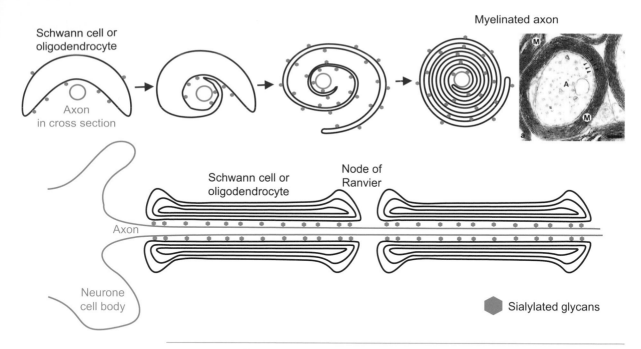

Figure 12.7 Myelin sheath formation by Schwann cells of the peripheral nervous system and oligodendrocytes of the central nervous system. Processes from the myelinating cells envelop and then spiral around axons, forming alternating layers of lipid bilayer membranes. [Electron micrograph of myelin reproduced from: Schachner, M. and Bartsch, U. (2000). Multiple functions of the myelin-associated glycoprotein MAG (siglec-4a) in formation and maintenance of myelin, *Glia* **29**, 154–165 with permission. Glycan portions of glycolipids are indicated schematically as hexagons.] A, axon; M, myelin.

However, although morphologically quite normal myelin sheaths are present in these mice, they suffer from progressive demyelination, indicating that the more complex glycolipids function in stabilization of the myelin sheath. In spite of the abundance of galactosphingolipids in myelin, sheath formation also occurs in knockout mice lacking the galactosyltransferase that initiates galactosphingolipid synthesis. Extra synthesis of glucosylceramide in the brains of these mice compensates for loss of the galactosylceramide and sulphatide, resulting in morphologically normal-looking myelin sheaths. However, these mice again show neurological defects, indicating that the myelin sheath is not fully functional.

The findings that myelin sheaths are formed in mice lacking either galactosylceramide glycolipids or the larger glucosylceramide lipids suggest that the general physical properties of the glycolipid head groups, rather than any specific sequences of sugars within these head groups, are a key to the myelination process. The head groups may help to maintain the proper spacing between the lipid bilayers. This function in myelin may represent a specialized form of a more general role of the glycolipid head groups in creating the meshwork of carbohydrates that forms a barrier around the surface of animal cells. The resulting modulation of the physical properties of the membrane is analogous to direct effects of glycosylation on protein structure and stability.

⊙ See Chapter 8 for more on the effects of glycosylation on protein structure and stability.

The diversity of glycosphingolipids, coupled with their abundance in neural tissues, suggests that they have the potential to serve as specific recognition markers that might play a crucial part in many recognition processes between cells in the nervous system. Antibodies against certain glycolipid epitopes selectively stain specific populations of neurons, which would be consistent with a role in the interaction of these cells with their neighbours during processes such as axon guidance. Lectins on adjacent cells or in the extracellular matrix could bind to some of these terminal structures. One such receptor, myelin-associated glycoprotein, which has a major role in the stabilization of myelin, is discussed in the next section. However, the number of known receptors does not come close to the number of glycans found on glycolipids, so there is unlikely to be a one-to-one correspondence of glycolipid markers and receptors. In contrast to the view that each different glycolipid head group may mediate a distinct function, an alternative extreme view is that classes of glycoconjugates such as the gangliosides might be essential for a properly functioning nervous system without individual glycans having distinct roles. As mice lacking specific subgroups of glycolipids are created and as their phenotypes are examined in more detail, the answer will probably turn out to be somewhere between these two extreme views: certain specific glycolipids mediate some specialized recognition events, but in many cases entire classes have overlapping and redundant functions.

Further evidence for the importance of glycolipids in development may come from the fact that, although several human diseases due to relatively common defects in enzymes for glycolipid breakdown are known, only one human genetic disorder of glycolipid synthesis has been seen, possibly because most defects in glycolipid synthesis would be lethal during embryogenesis. A defect in ganglioside biosynthesis, due to a mutation in the gene for the sialyltransferase (LacCer α2-3 sialyltransferase or sialyltransferase 9) that adds sialic acid to lactosylceramide to form the ganglioside GM3, has been characterized in an Old Amish family. The nonsense mutation is in the region between the two sialylmotifs and results in a non-functional protein. Formation of GM3 is the first step in the synthesis of most complex gangliosides so deficiency of LacCer α2-3 sialyltransferase affects synthesis of many glycolipids, not just GM3. Biochemical analysis of plasma and fibroblasts from the affected individuals shows an almost complete lack of GM3 and other gangliosides normally synthesized from GM3. Children who are homozygous for the mutated gene develop epilepsy in the first year of life and have severe developmental defects. This human disease emphasizes the importance of normal ganglioside biosynthesis for brain development, but provides only limited insight into the exact functions of glycolipids in development.

Glycobiology and disease
Glycolipid synthesis and early on-set epilepsy

⊘ See section 4.5 for more on glycolipid storage disorders.

⊘ See section 5.1 for more on sialyltransferases.

12.6   Myelin-associated glycoprotein has roles in development of the central and peripheral nervous systems

Unlike all the other siglecs, which are found on cells of the immune system, myelin-associated glycoprotein (MAG) is expressed in the nervous system, on the surface of myelin-forming Schwann cells and oligodendrocytes. MAG comes in two alternatively

⊘ See sections 9.9 and 9.10 for more on siglecs.

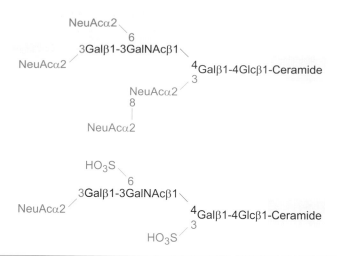

Figure 12.8 Glycolipid ligands for myelin-associated glycoprotein (MAG). Gangliosides (sialylated glycosphingolipids) bearing multiple negative charges, in the form of sialic acid residues and sulphate groups, have been implicated in the function of MAG.

spliced forms, one with a longer cytoplasmic tail (L-MAG) and one with a short tail (S-MAG). L-MAG functions in the central nervous system in initiation of myelination. In contrast, S-MAG helps to maintain the mature peripheral nervous system by stabilizing axon–myelin interactions and inhibiting neurite outgrowth. All of these roles involve the sialic acid-binding activity of the terminal domain. Like sialoadhesin, MAG binds to ligands containing NeuAc in α2-3 linkage to galactose. However, MAG discriminates more strongly between ligands and does not bind glycans containing NeuAc in α2-6 linkage. In addition, there is evidence that the binding site in MAG may be more extended than the site in sialoadhesin. MAG binds particularly well to complex glycolipids that bear terminal sialic acid residues alongside additional negative charges in the form of sialic acid or sulphate substituents (Figure 12.8).

The natural ligands for MAG are not known with certainty, but it can bind to glycoproteins as well as glycolipids. *In vitro* assays demonstrating high-affinity binding to complex glycolipids, combined with the presence of MAG in the myelin sheath around axons, are consistent with roles in the formation and stabilization of the myelin sheath. The peripheral nervous systems of knockout mice lacking MAG show decreased levels of myelination. The myelination process appears to be essentially normal in the central nervous system of these mice, but there is a deterioration of the interaction between the axon and the inner layer of the myelin as they age. The resulting separation between axon and myelin leads to neurological defects. The phenotype of the knockout mice lacking MAG is very similar to that of mice in which the GalNAc-transferase necessary for the formation of extended gangliosides has been deleted, suggesting that MAG interactions with gangliosides may be important in the stabilization of myelin. Mice expressing only S-MAG display the central

Box 12.2   233

nervous system defects but not those in the peripheral nervous system. Inhibition of neurite outgrowth mediated by MAG may be an important reason that nervous system regeneration is inefficient and it would be of clinical interest to be able to overcome this inhibition in cases of damage to the nervous system. *In vitro* assays using rat cerebellar neurons cultured on MAG-containing myelin show that synthetic glycosides containing glycans that bind tightly to MAG, NeuAcα2-3Galβ1-3(NeuAc α2-6)GalNAc- and NeuAcα2-3Galβ1-3GalNAc- (Figure 12.8) can enhance outgrowth of neurons. Such glycans might eventually be used therapeutically, possibly together with sialidases and chondroitinases (see Box 12.2), in aiding regeneration of the nervous system following injury.

 BOX 12.2   **Glycotherapeutics:** *Chondroitinase and sialidase treatments facilitate regeneration in the central nervous system*

In mammals, neurons of the central nervous system fail to regenerate following injury. This lack of regeneration following spinal cord injury is a major cause of permanent paralysis. Both motor and sensory connections in the peripheral nervous system can be re-established after injury, suggesting that regeneration in the central nervous system might be possible if inhibitory factors could be identified and neutralized. One source of inhibitory factors is disrupted myelin, which contains components such as myelin-associated glycoprotein (MAG). Because the axon regrowth inhibitory activity of MAG is dependent on its interaction with sialylated ganglioside ligands (see section 12.6), it has been suggested that destroying such ligands might relieve the growth inhibition. In animal model studies to test the potential of this approach, injection of bacterial sialidase into a site of contusion injury in the spinal cord stimulates axonal growth and recovery of function.

Chondroitin sulphate-containing proteoglycans in the brain extracellular matrix are additional inhibitors of nerve growth. Study of neurons in culture reveals that axons change their direction of growth when they contact chondroitin sulphate. In the embryo, production of chondroitin sulphate proteoglycans in specific regions is believed to create barriers to axon extension in specific areas, although in other areas the matrix facilitates outgrowth and guides axons. The adult matrix is largely inhibitory.

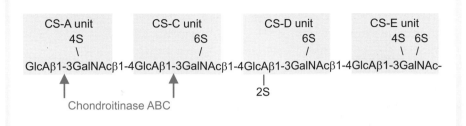

Multiple matrix proteoglycans in the brain, including aggrecan and a related family of proteins (see Chapter 3), bear chondroitin sulphate chains. Like heparan sulphate, chondroitin sulphate is modified during biosynthesis and contains regions with differing degrees of sulphation, with regions that consist of CS-D and CS-E units being sulphate-rich. Chondroitin sulphate fragments that differ in sulphation can have either stimulatory or inhibitory effects on axon growth in culture. By analogy to heparan sulphate, it seems likely that specific forms of chondroitin sulphate initiate or modify cellular signalling events, although the pathways involved and the specific requirements for chondroitin sulphation have not been defined. In any case, the ability of different astrocyte cell lines to inhibit axon growth correlates with the amount of certain chondroitin sulphate proteoglycans that they express. Also, the accumulation of astrocytes and proteoglycans produced at the site of spinal cord injury, known as the glial scar, is rich in inhibitory proteoglycans due to extra core protein synthesis and increased sulphation on chondroitin sulphate. The scar prevents axon regrowth through the area of damage.

The role of chondroitin sulphate as an inhibitor of axon extension suggests that the inhibition might be eliminated by decreasing the levels of proteoglycans at the site of injury. Because inhibition is mediated by the glycosaminoglycan chains, enzymes that degrade chondroitin sulphate have been investigated as possible therapeutic agents. Bacterial chondroitinases degrade several of the linkages that occur in chondroitin sulphate chains and tests in cell culture confirm that nerve growth on chondroitin sulphate-containing matrix is enhanced following treatment with chondroitinase. In preclinical tests, the effect of chondroitinase *in vivo* has been investigated following injection at sites of spinal cord injuries created by cutting or compressing nerves. Encouraging evidence of enhanced nerve regeneration, leading to functional re-innervation of target tissues, has been obtained.

Essay topics

- Describe the families of proteoglycans found in the brain.
- Discuss the evidence that chondroitin sulphate can modulate axon growth in the central nervous system.

Lead references

- Bradbury, E.J., Moon, L.D.F., Popat, R.J., King, V.R., Bennett, G.S., Patel, P.N., Fawcett, J.W., and McMahon, S.B. (2002). Chondroitinase ABC promotes functional recovery after spinal cord injury, *Nature* **416**, 636–640.

- Caggiano, A.O., Zimber, M.P., Ganguly, A., Blight, A.R., and Gruskin, E.A. (2005). Chondroitinase ABCI improves locomotion and bladder function following contusion injury of the rat spinal column, *Journal of Neurotrauma* **22**, 226–239.

- Mountney, A., Zahner, M.R., Lorenzini, I., Oudega, M.R., Schramm, L.P., and Schnaar, R.L. (2006). Sialidase enhances recovery from spinal cord contusion injury, *Proceedings of the National Academy of Sciences U.S.A.* **107**, 11561–11566.

- Oohira, A., Matsui, F., Tokita, Y., Yamauchi, S., and Aono, S. (2000). Molecular interactions of neuronal chondroitin sulfate proteoglycans in brain development, *Archives of Biochemistry and Biophysics.* **374**, 24–34.

- Silver, J. and Miller, J.H. (2004). Regeneration beyond the glial scar, *Nature Reviews Neuroscience* **5**, 146–156.

- Mountney, A., Zahner, M.R., Lorenzini, I., Oudega, M.R., Schramm, L.P., and Schnaar, R.L. (2006). Sialidase enhances recovery from spinal cord contusion injury, *Proceedings of the National Academy of Sciences U.S.A.* **107**, 11561–11566.

12.7   Polysialylation of neural cell adhesion molecule prevents cell adhesion during development

A major role of membrane protein glycosylation is to present various terminal structures that can be bound by lectins. However, glycosylation can also modulate adhesion in other ways. One of the best studied examples of such modulation is the effect of polysialic acid addition to cell surface glycans. Chains consisting of eight to 100 or more *N*-acetylneuraminic acid (NeuAc) residues in α2-8 linkage can be added specifically to the neural cell adhesion molecule (NCAM), which mediates adhesion between cell surfaces by forming a homotypic interaction. Polysialic acid chains added to N-linked glycans on NCAM prevent this homotypic interaction and are therefore anti-adhesive. The polysialic acid chains cause repulsion due to their bulk and their negative charges. Repulsion between membrane surfaces bearing polysialylated NCAM can inhibit adhesion functions of other cell surface receptors as well as those mediated by NCAM itself. Addition of polysialic acid is correlated with the ability of cells to move or change shape. Cell migration and axon development during brain development are thus associated with polysialylation of NCAM. The level of polysialylation of NCAM decreases dramatically as brain development slows, reflecting an increasing number of stable interactions between neurons as the wiring circuitry of the adult brain is established.

12.8   Changes in glycosylation occur during cell differentiation in the immune system

T lymphocytes provide a good example of a cell type where expression of cell surface glycans is regulated during differentiation. T cell development is associated with changes in expression of O-linked glycans on cell surface receptors. In the thymus, thymocytes, which are the precursors of T cells, first express both of the T cell co-receptors CD4 and CD8. These double positive ($CD4^+CD8^+$) thymocytes undergo either negative selection leading to cell death by apoptosis, or positive selection to become cells expressing just CD4 or just CD8 (Figure 12.9). During maturation of T cells in the thymus, expression of core 2 type O-glycans changes due to decreased expression of the core 2 β-1, 6-*N*-acetylglucosaminyltransferase (core 2 GnT) needed to add GlcNAc in 1-6 linkage to the GalNAc residue of core 1 type O-glycans. Thus, core 2 glycans are found attached to proteins such as CD43 at the surface of immature, double positive ($CD4^+CD8^+$) thymocytes undergoing selection in the cortical part of the thymus, but not on the surface proteins of mature single positive ($CD4^+$ or $CD8^+$) T cells in the thymus medulla. Expression of core 2 GnT and core 2 glycans correlates with the developmental stage where negative selection of self-reactive or non-reactive thymocytes occurs by apoptosis. Core 2 glycans can be extended on the 1-6 branch with polylactosamine to provide

⊕ See section 3.1 for more on core 1 and core 2 O-glycans.

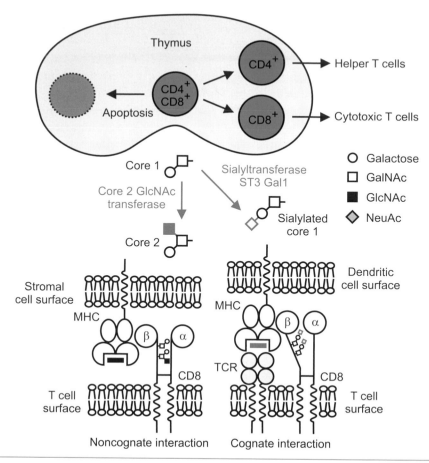

Figure 12.9 Changes in CD8 glycosylation during T cell maturation. As thymocytes mature from CD4⁺CD8⁺ double positive cells to CD8⁺ single positive cells, the expression of core 2 GalNAc transferase declines and expression of sialyltransferase ST3 Gal1 increases, leading to a change in the O-linked glycans on CD8. Sialylated core 1 structures at key sites in the stalk of CD8 change the disposition of the terminal domains, so that instead of interacting non-specifically with MHC molecules on stromal cells, they now facilitate interaction between the MHC molecules and T-cell receptors that specifically recognize the bound peptide. MHC, major histocompatibility complex; TCR, T cell receptor.

⊙ See section 9.12 for more on galectins and apoptosis.

ligands for galectin-1, which triggers apoptosis through cross-linking of cell surface receptors.

Changes in glycans on cell surface receptors are also associated with the positive selection of CD8⁺ cytotoxic T cells. The transition from double positive immature thymocytes to mature thymocytes expressing only CD8 is associated with addition of sialic acid to core 1 O-glycans on CD8. The sialyltransferase, ST3Gal-1, which adds sialic acid to galactose of core 1 glycans, is expressed by mature single positive thymocytes but not by double positive thymocytes. Knockout mice lacking ST3Gal-1 have reduced numbers of CD8⁺ T cells providing evidence for the importance of sialylation in positive selection. The presence of sialic acid on the O-glycans of CD8 modulates the strength of interaction between CD8 and MHC class I molecules. In the thymus, MHC molecules are present on the stromal cells and interactions between the thymocytes and MHC molecules are critical in T cell selection. CD8 usually acts as a co-receptor for the interaction of the T-cell receptor with MHC class I molecules presenting peptide recognized by the T-cell receptor. However, CD8 can also bind to MHC class I molecules that are not recognized by the T-cell receptor. Such non-cognate interactions are stronger between MHC class I molecules and non-sialylated CD8 on developing double-positive thymocytes than between MHC class I molecules and sialylated CD8 on mature CD8$^+$ thymocytes. Non-cognate interactions between CD8 and MHC class I molecules might be important for enhancing interaction between stromal cells and thymocytes during positive selection. Developmentally regulated sialylation provides a means of switching off this interaction following maturation of single positive thymocytes.

The core 1 O-glycans that are sialylated on CD8 are attached to several threonine residues in the stalk region of the β-chain polypeptide that projects the immunoglobulin-like binding domain away from the cell surface (Figure 12.9). Exactly how sialylation of these glycans reduces the affinity of CD8 for MHC class I molecules is not clear. Sialylation of the β-chain stalk region glycans might induce a conformational change that alters the orientation of the immunoglobulin-like domains in the CD8αβ heterodimer. Alternatively, association of the α and β immunoglobulin-like domains might be reduced due to the presence of the larger, negatively charged glycans. Either of these changes could decrease the ability of the CD8 immunoglobulin-like domains to bind to MHC class I molecules.

SUMMARY

The genetic and biochemical results described in this chapter provide compelling evidence that many types of glycosylation play important roles in establishing the basic body plan of vertebrates and invertebrates, in development of organs, and in maturation of the nervous and immune systems. In several cases, specific glycan structures needed for these processes have been identified, but in most cases we lack a clear understanding of how these glycans exert their effects. A major goal for the future will be to define how these glycans work, either through binding to specific receptors or by modulating the behaviour of molecules at the surfaces of interacting cells.

KEY REFERENCES

Coetzee, T., Fujita, N., Dupree, J., Shi, R., Blight, A., Suzuki, K., Suzuki, K., and Popko, B. (1996). Myelination in the absence of galactocerebroside and sulfatide: normal structure with abnormal function and regional instability, *Cell* **86**, 209–219. This paper describes the phenotype of knockout mice that cannot synthesize galactocerebroside or sulphatide.

Collins, B.E., Ito, H., Sawada, N., Ishida, H., Kiso, M., and Schnaar, R.L. (1999). Enhanced binding of the neural siglecs, myelin-associated glycoprotein and Schwann cell myelin protein, to Chol-1 (alpha-series) gangliosides and novel sulfated Chol-1 analogs, *Journal of Biological Chemistry* **274**, 37637–37643. Experiments that define the extended carbohydrate binding specificity of MAG are described.

Gascoigne, N.R.J. (2002). T-cell differentiation: MHC class I's sweet tooth lost on maturity, *Current Biology* **12**, R99–R101. A new model is proposed for how cell surface glycans modulate lymphocyte activation.

Jafar-Nejad, H., Leonardi, J., and Fernandez-Valdivia, R. (2010). Role of glycans and glycosyltransferases in the regulation of Notch signaling, *Glycobiology* **20**, 931–949. A review of the effects of multiple types of glycosylation on Notch is presented.

Kreuger, J., Salmivirta, M., Sturiale, L., Gimenez-Gallego, G., and Lindahl, U. (2001). Sequence analysis of heparan sulphate epitopes with graded affinities for fibroblast growth factors 1 and 2, *Journal of Biological Chemistry* **276**, 30744–30752. Key experiments defining the sequence of heparan sulphate saccharides that bind to growth factors are presented.

Lander, A.D. and Selleck, S.B. (2000). The elusive functions of proteoglycans: *in vivo* veritas, *Journal of Cell Biology* **148**, 227–232. *Drosophila* and mammalian mutations affecting proteoglycans are compared in a thoughtful review.

Nawroth, R., van Zante, A., Cervantes, S., McManus, M., Hebrok, M., and Rosen, S.D. (2007). Extracellular sulfatases, elements of the Wnt signalling pathway, positively regulate growth and tumorigenicity in human pancreatic cancer cells, *PLOS One* **2**, e392. This paper describes how sulphatases may control the activity of growth factors by remodelling glycosaminoglycans.

Park, P.W., Reizes, O., and Bernfield, M. (2000). Cell surface heparan sulfate proteoglycans: selective regulators of ligand-receptor encounters, *Journal of Biological Chemistry* **275**, 29923–29926. A short review is provided covering details of interactions between heparan sulphate proteoglycans and their protein ligands, and the cellular processes modulated by these interactions.

Pellegrini, L. (2001). Role of heparan sulfate in fibroblast growth factor signalling: a structural view, *Current Opinion in Structural Biology* **11**, 629–634. Insights gained from crystal structures of growth factors in complex with heparin are reviewed.

Rapraeger, A.C. (2000). Syndecan-regulated receptor signaling, *Journal of Cell Biology* **149**, 995–997. A short commentary reviewing evidence for roles of the cell–surface proteoglycan syndecan in modulating cell-matrix adhesion and signalling is presented.

Rutishauser, U. (2008). Polysialic acid in the plasticity of the developing and adult vertebrate nervous systems, *Nature Reviews Neuroscience* **9**, 26–35. A detailed review of the *in vitro* and *in vivo* evidence for the role of polysialic acid is presented.

Schnaar, R.L. (2010). Brain gangliosides in axon-myelin stability and axon regeneration, *FEBS Letters* **584**, 1741–1747. The evidence for roles of gangliosides in maintenance of myelin and in nerve regeneration is reviewed.

Schnaar, R.L. and Lopez, P.H.H. (2009). Myelin-associated glycoprotein and its axonal receptors, *Journal of Neuroscience Research* **87**, 3267–3276. The functions of MAG in the nervous system are reviewed in detail.

Wallis, G.A. (1995). Cartilage disorders. The importance of being sulphated, *Current Biology* **5**, 225–227. This is a short commentary on the finding that mutations in the gene for a sulphate transporter underlie a cartilage disorder, which shows the importance of sulphation of glycosaminoglycans.

Yamashita, T., Wada, R., Sasaki, T., Deng, C., Bierfreund, U., Sandhoff, K., and Proia, P.L. (1999). A vital role for glycosphingolipid synthesis during development and differentiation, *Proceedings of the National Academy of Sciences U.S.A.* **96**, 9142–9147. The phenotype of mice in which the gene for glucosylceramide synthase has been knocked out, preventing synthesis of glycosphingolipids, is described.

QUESTIONS

12.1   Discuss how glycosylation is important in signalling pathways.

12.2   Compare the findings of the following two studies which investigated the specificity of binding of heparan sulphate proteoglycans to two different families of growth factors.

References: Kreuger, J., Jemth, P., Sanders-Lindberg, E., Eliahu, L., Ron, D., Basilico, C., Salmivirta, M., and Lindahl, U. (2005). Fibroblast growth factors share binding sites in heparan sulphate, *Biochemical Journal* **389**, 145–150.

Pankonin, M.S., Gallagher, J.T., and Loeb, J.A. (2005). Specific structural features of heparan sulfate proteoglycans potentiate neuregulin-1 signalling, *Journal of Biological Chemistry* **280**, 383–388.

12.3   Discuss the importance of glycan–protein interactions in development of the nervous system.

12.4   Knockout mice lacking enzymes involved in glycan synthesis have provided insights into the importance of glycans in development, but sometimes knockout experiments produce surprising results. Explain how the unexpectedly mild phenotype of mice lacking Golgi mannosidase II lead to the discovery of an alternate pathway for synthesis of N-linked glycans.

Reference: Chui, D., Oh-Eda, M., Liao, Y.F., Panneerselvam, K., Lal, A., Marek, K.W., Freeze, H.H., Moreman, K.W., Fukuda, M.N., and Marth, J.D. (1997). Alpha-mannosidase-II deficiency results in dyserythropoiesis and unveils an alternate pathway in oligosaccharide biosynthesis, *Cell* **90**, 157–167.

12.5   The mechanism by which the extended glycans created by the action of the Fringe glycosyltransferase modify the interactions of Notch with Delta and Serrate is not yet known. Suggest some possible mechanisms. One possible mechanism is proposed in the following paper. Assess the evidence to support this proposal.

Reference: Xu, A., Lei, L., and Irvine, K.D. (2005). Regions of *Drosophila* Notch that contribute to ligand binding and the modulatory influence of Fringe, *Journal of Biological Chemistry* **280**, 30158–30165.

13 Glycosylation and disease

LEARNING OBJECTIVES

By the end of this chapter, students should understand:

1 The molecular basis of molecular dystrophy and other diseases that result from defects in the glycan biosynthetic machinery

2 Normal and aberrant immune responses to carbohydrates

3 Issues associated with production of recombinant glycoproteins as therapeutics

4 The nature and possible effects of glycosylation changes associated with cancer and other diseases

Examples of diseases associated with aberrant glycosylation or failure of carbohydrate recognition systems have been cited at many points in this book. Much of our understanding of the biological roles of glycans and their receptors derives from naturally occurring diseases and disease phenotypes created in knockout mice. The purpose of this chapter is to highlight some additional ways in which glycosylation is associated with diseases. Like many of the examples discussed earlier, the first two diseases described here are genetic in origin. Further examples involve chemical reactions of sugars, immunological responses to them, and diseases in which glycosylation changes are correlated with disease phenotype, but where a causal connection is not yet clear.

13.1 Mutations in enzymes for synthesis of N-linked glycans cause congenital disorders of glycosylation

The molecular bases for several distinct congenital disorders of glycosylation (CDGs) have been identified. These inherited conditions have usually been recognized because of changes in glycosylation of serum glycoproteins, such as transferrin, in afflicted patients. They were at one time referred to as carbohydrate-deficient glycoprotein syndromes and are quite rare, having been detected in only a few hundred individuals worldwide. The syndromes can be categorized into two general types. The first group consists of syndromes in which essentially normal N-linked glycans are attached to serum glycoproteins, but the frequency with which potential N-glycosylation sites are utilized is reduced compared with glycoproteins from control serum. In the second type of syndrome, all or most of the usual glycosylation sites are utilized, but the attached glycans are reduced in size and complexity.

CDGs of the first type result from reduction in the pool of dolichol-linked high-mannose precursors that serve as donors in the initial step of N-glycosylation. The molecular defects in two forms of this syndrome, CDG-1a and CDG-1b, lie in the biosynthetic pathways for generation of dolichol-oligosaccharide (Figure 13.1). The most common defect (CDG-1a) is in phosphomannomutase, the enzyme that normally catalyses isomerization of mannose 6-phosphate to mannose 1-phosphate. A less common form of the disease (CDG-1b) results from defects in the phosphomannose isomerase gene which generates mannose 6-phosphate from fructose 6-phosphate. It is not clear how the resulting hypoglycosylation leads to the range of clinical symptoms, including developmental defects and loss of muscle tone, associated with these diseases. However, it seems likely that defects in cell surface and matrix glycoproteins may well be involved. In the case of the relatively rare CDG-1b, a bypass route to synthesis of GDP-mannose is available, because mannose can be directly phosphorylated to generate mannose 6-phosphate. High levels of mannose in the diet are sufficient to restore normal levels of the dolichol donor and thus prevent hypoglycosylation of serum glycoproteins. Unfortunately, no such remedies are available for other CDGs.

The second group of CDGs result from defects further along in the glycosylation pathway. One form of the disease is caused by the absence of GlcNAc transferase II, which is responsible for the addition of GlcNAc to the 1–6 arm of complex glycans. In this syndrome, serum glycoproteins are rich in structures in which the 1–3 arm is elaborated as usual, but the 1–6 arm is truncated. As in the case of the type I CDGs, it is difficult to explain the observed phenotypes based on the molecular defects. For this reason,

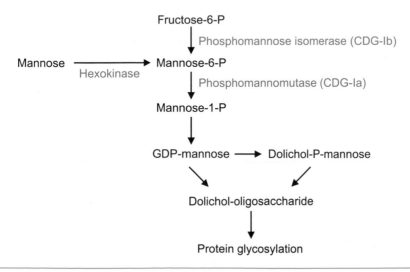

Figure 13.1 Pathway for mannose incorporation into glycoproteins. Mutations in phosphomannose isomerase or phosphomannomutase lead to CDG types Ib and Ia. CDG-Ib can be treated with mannose because hexokinase allows the missing enzyme to be bypassed. GDP, guanosine diphosphate.

analysis of these syndromes provides only limited insights into the exact functions of glycans, but does confirm that complex glycans have important roles in development.

13.2 Abnormal expression of a glycosyltransferase causes a blood clotting defect

Von Willebrand factor deficiency is a relatively common bleeding disorder, affecting up to one in a hundred individuals. The most common form of the disease is inherited in simple Mendelian fashion, suggesting that it results from an alteration at a single genetic locus. Importantly, the mutant gene is effectively dominant. Serum levels of von Willebrand factor in individuals with the deficiency are reduced two- to five-fold compared with normal, but a small amount of the factor is present and the activity of this protein is normal. Because von Willebrand factor is essential for initiation of the blood-clotting cascade, this decreased serum level results in excessive bleeding following wounding. In a mouse model for the human disease, the affected gene maps to a region of chromosome 11, while the gene encoding von Willebrand factor is located on chromosome 6.

One gene in the relevant region of mouse chromosome 11 encodes a GalNAc transferase. No mutations in the coding region of this gene are found in the affected mice, but a change in the promoter region causes the protein to be expressed in endothelial cells rather than in epithelial cells (Figure 13.2). Because von Willebrand factor is made in endothelial cells, it would not normally encounter the GalNAc-transferase. Mislocalization of the transferase results in co-expression with von Willebrand factor in endothelial cells, leading to the addition of terminal GalNAc

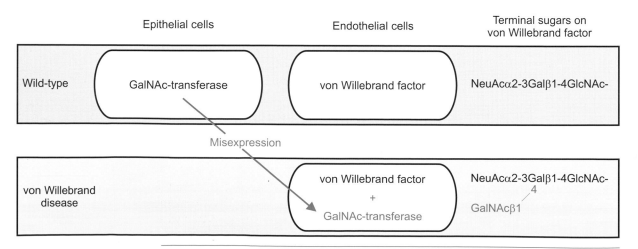

Figure 13.2 Altered glycosylation of von Willebrand factor resulting from misexpression of GalNAc-transferase. The presence of the transferase in endothelial cells leads to addition of terminal GalNAc residues to von Willebrand factor, making it a target for clearance by the asialoglycoprotein receptor.

residues to the von Willebrand factor. The terminal GalNAc residues on the secreted protein make it a ligand for the asialoglycoprotein receptor, which clears the abnormally glycosylated factor from the circulation. The activity of the transferase explains the dominant phenotype, because all von Willebrand factor is subject to addition of GalNAc as long as the transferase is expressed in endothelial cells even if only from a single mutant gene.

⊙ See section 10.7 for more on the asialoglycoprotein receptor.

These results demonstrate that changes in glycosylation can have remarkably subtle and indirect effects. Although changes in glycosylation might be global, phenotypes associated with such changes are often most evident for one or a few proteins. The mislocalization of GalNAc transferase probably causes aberrant glycosylation of a number of different proteins in endothelial cells, but the major physiological effect observed is the bleeding disorder.

13.3   Chemical glycation of proteins occurs in diabetes

Throughout this book, it has been emphasized that addition of glycans to glycoproteins requires the action of specific glycosyltransferases. Although this principle holds true for the normal forms of glycosylation, a direct chemical reaction of sugars with proteins does occur. The process is referred to as glycation. The reactions are generally inefficient, so they are usually observed only under special circumstances, such as when glucose levels become very high in diabetic patients or in proteins that have very long half-lives.

The chemical details of the reactions that occur during glycation are not fully understood, but they are known to involve the small amount of non-cyclized glucose that is in equilibrium with the usual pyranose form. The free carbonyl group present in this open chain structure is able to form a Schiff base with an amino group on a protein (Figure 13.3). Migration of the double bond, in the Amadori reaction, leads to a stable product that accumulates spontaneously when proteins are incubated with glucose. Poor control of glucose levels in diabetic patients results in spikes in blood glucose levels that accelerate the rate of accumulation of such glycated proteins. For this reason, the degree to which a particular regimen of insulin delivery matches the dietary glucose load can be judged by measuring the levels of glycated haemoglobin or serum albumin. Glycated products of serum proteins can be cleared from circulation by specific receptors. These receptors appear to be distinct from the animal lectins that clear normal glycoproteins, reflecting the chemically very distinct nature of the glycation products compared with N- and O-linked glycans.

The Amadori product can also react further in a number of ways. For example, reaction with an additional amino group leads to the cross-linking of proteins (Figure 13.3). Glycation-mediated cross-linking of collagen, a long-lived protein, underlies kidney and vascular damage that is often a complication associated with diabetes. Similar reactions occur over an extended time in all individuals and the products of these reactions are referred to as advanced glycation end products (AGEs). Some products of the reactions following Schiff base formation and the Amadori reaction contain multiple conjugated double bonds and thus absorb light

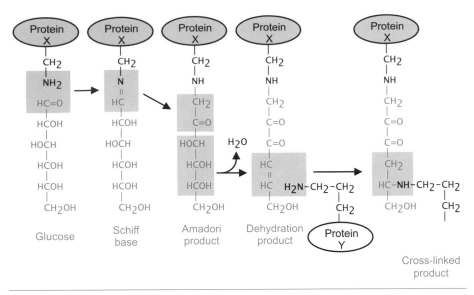

Figure 13.3 Glycation reactions leading to protein cross-linking and formation of advanced glycation end products. Glucose, when present in serum at persistently high concentrations, reacts with an amino group on a lysine side chain of a protein. Following rearrangement, further reaction with a lysine side chain on another protein occurs, resulting in cross-linking of the proteins.

in the visible region. The formation of such chromophores is particularly evident in normally transparent tissues such as the lens, which becomes yellowed as a result. The crystallin proteins that make up the lens do not turn over, so damage sustained in this way accumulates continuously with age.

13.4   Antibodies to carbohydrates can cause disease

➲ See section 2.7 for more on ABO blood groups.

The antigenic nature of carbohydrates is evident from the ABO blood group system. Individuals who do not synthesize either of the terminal sugar structures that form the A and B epitopes develop antibodies to these antigens, because similar glycans are found in various food substances. Individuals who display the antigens on their own cells from birth are tolerant to them and do not generate an antibody response. The presence of these glycans and the antibodies to them vary from individual to individual, because of genetic polymorphism within the human population. The antibody responses are harmless under normal circumstances, because the antigens are only encountered in the digestive system. Problems arise when blood cells bearing one type of glycan are transfused from one individual to another who does not make this antigen and therefore has developed antibodies to it.

Anticarbohydrate antibodies that result from differences in the carbohydrate structures found on cells of humans and other animal species can also cause disease. Most vertebrates, including Old World primates, contain a gene for an

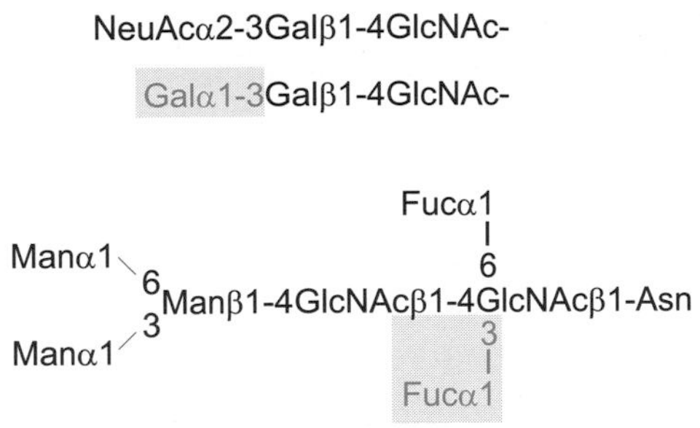

NeuAcα2-3Galβ1-4GlcNAc-

Galα1-3Galβ1-4GlcNAc-

```
                              Fucα1
                               |
   Manα1\                       6
         \6                     6
          Manβ1-4GlcNAcβ1-4GlcNAcβ1-Asn
         /3                     3
   Manα1/                       |
                              Fucα1
```

Figure 13.4 Antigenic carbohydrate structures derived from non-human tissues with novel sugar linkages highlighted. α-Galactosyl residues are present in most mammals, including non-human primates, while α1-3-linked core fucose residues are present on many insect glycoproteins.

α-galactosyltransferase that is able to cap glycan structures by forming a Galα1-3Gal linkage (Figure 13.4). This terminal structure provides an alternative to the sialylation of N-linked glycans, but it also can be attached to other glycoprotein and glycolipid glycans. Although the α-galactosyltransferase gene is present in humans and New World monkeys, it is not expressed and thus the Galα1-3Gal epitope is not present on endogenous glycans. As in the case of the ABO blood group substances, environmental exposure to similar epitopes on food substances and microbes leads to the production of antibodies reactive with terminal α-galactosyl residues in most humans. These antibodies represent a major obstacle to xenotransplantation of organs, because organs from preferred donors such as pigs are subject to hyperacute rejection as they quickly become coated with antibodies to the terminal α-galactosyl residues. Although absorption of antibodies from the serum of potential transplant recipients alleviates the immediate problem, it seems likely that animals lacking the transferase activity will have to be created as sources of organs compatible with the human immune response.

Some carbohydrate xenoantigens cause illness because of the nature of the immune response to them. Fucose residues such as those found in the ABO blood group substances are referred to as outer arm fucose residues to distinguish them from fucose residues that are often attached to the GlcNAc residues in the N-linked core. In vertebrates, such core fucose residues are attached in α1-6 linkage to the inner, asparagine-linked GlcNAc residue (Figure 13.4). This modification is also found in plant and insect glycoproteins, but in these species fucose can also be added to the inner GlcNAc residue in α1-3 linkage, sometimes alone and sometimes in the presence of α1-6-linked fucose. Exposure to these glycoproteins, probably through the diet, can lead to antibody production, which in some individuals takes the form of an IgE response. Because binding of IgE molecules to antigen often results in an allergic response, such individuals are allergic to glycans containing the difucosylated structure. A common route of exposure to such glycans is in the form of bee stings, as N-linked glycoproteins in bee venom are heavily fucosylated in the core region. The deposition of venom under the skin leads to a local reaction in all individuals, because the protein melittin acts to stimulate nerves by depolarizing them.

However, in individuals with circulating IgE molecules that recognize the difucosylated epitope, a systemic anaphylactic response often occurs, in which IgE-stimulated release of histamines and other compounds from mast cells can cause fatal constrictions of the airways in the lung.

13.5   Producing glycoproteins to treat many diseases is a challenge for biotechnology

Among the rapidly increasing number of therapeutic glycoproteins being produced are many designed to replace ones that are missing as a result of disease conditions, factors to stimulate cell growth and regeneration, and monoclonal antibodies such as those used to target tumour cells. Some of these glycoproteins, such as enzymes for replacement therapy of lysosomal storage diseases, are used in a relatively small number of patients, but others like recombinant erythropoietin as a treatment for anaemia, clotting factors for treatment of blood coagulation disorders, and monoclonal antibodies have potentially very large markets. Proper glycosylation of these proteins is essential both because of the physical properties such as solubility and resistance to proteases that the sugars confer on the proteins (Chapter 8) and because the presence of the wrong glycans can target the glycoproteins for rapid clearance by endocytic glycan-binding receptors such as the mannose receptor and the asialoglycoprotein receptor.

Many therapeutic glycoproteins approved for use in humans are currently produced in Chinese hamster ovary (CHO) and baby hamster kidney (BHK) cells. It might seem that the most efficient way to mimic natural human glycosylation would be to use human cells, but there are safety concerns with this approach because of the potential for contamination with human viruses. Although hamster cells produce glycoproteins with complex N-linked glycans, they produce glycans that are typically less branched, bear more core fucose, and have more α2-3 and less α2-6-linked sialic acid than typical human glycoproteins. They also produce a heterogeneous mixture of different glycans and this can be a significant issue because of potential variation in the distribution of glycoforms between batches, because some glycoforms are likely to be more active or have longer half-lives than others. Mutant cell lines may be used to produce more uniform glycosylation, although the forms produced may not be ideal.

Systems that can be used more easily on industrial scales include insect cells, which produce mostly short paucimannose chains, and yeast, which produces N-linked glycans with extended mannan-type chains that differ in various species (Figure 13.5). Two general approaches are being investigated for employing these cells in recombinant glycoprotein production: glycan remodelling *in vitro* or engineering of the glycosylation machinery. The paucimannose structures from insect cells can be used directly as substrates for addition of terminal sugars using *in vitro* glycosyltransferase reactions. Mannosidases can be used to reduce yeast glycoproteins to a paucimannose state that can be similarly modified. Alternatively,

⊙ See **Box 4.1** for more on enzyme replacement therapy.

⊙ See **Box 10.2** for more on erythropoietin.

⊙ See **sections 10.7** and 10.8 for more on glycoprotein clearance.

⊙ See **section 5.7** for more on mutant CHO cells.

⊙ See **section 6.11** for more on glycosyltransferases as synthetic tools.

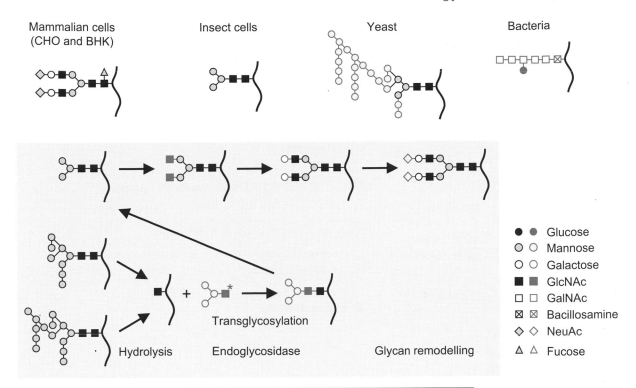

Figure 13.5 N-linked glycosylation produced in various systems for expression of recombinant glycoproteins. Regions of the structures which do not correspond to mature, complex glycans in humans are highlighted for each system. Methods for remodelling of glycoproteins include core modification employing the transglycosylation activity of endoglycosidases starting with oxazoline donors and terminal modifications with glycosyltransferases. Oxazoline group is indicated as *.

endoglycosidases can be used to replace core structures. In the *in vivo* approach, yeast have been engineered to incorporate branching GlcNAc transferases and galactosyltransferases, as well as the enzymes and transporters needed to make their substrates accessible, in order to produce nearly complete mammalian-type glycans directly. Similar engineering of insect cells is also being attempted. The successful transfer of the glycosylation machinery of campylobacter into *Escherichia coli* also opens up the possibility of making glycoproteins in bacteria, although substantial modification to the specificities of several of the steps in the process will be needed. Thus, the field of glycosylation engineering is evolving rapidly.

⬤ See section 11.2 for more on glycosylation in bacteria.

13.6 Genetic changes that modify unusual O-linked glycans can cause muscular dystrophy

Recent studies on congenital (inherited) forms of muscular dystrophy have led to increased interest in a novel class of O-linked glycans in which mannose is linked to

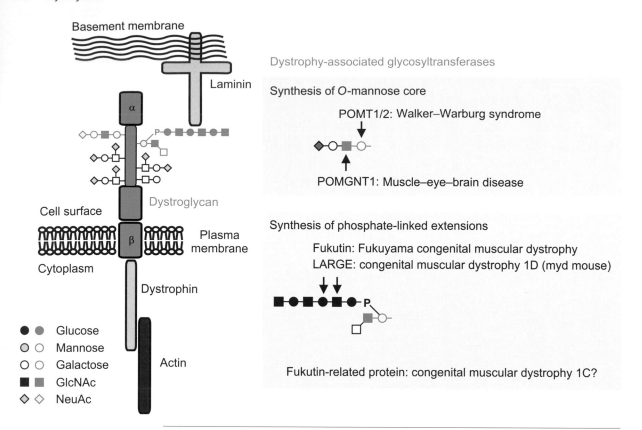

Figure 13.6 Aberrant glycosylation of α-dystroglycan in congenital muscular dystrophies. The role of dystroglycan is to link the intracellular cytoskeleton with the extracellular matrix. Of the five glycosyltransferases that are mutated in various forms of muscular dystrophy, two are required for synthesis of the core of mannose-linked O-glycans. The LARGE protein and possibly Fukutin are hypothesized to make a repeating polymer that is attached to the O-mannose through a phosphodiester linkage to the 6-hydroxyl group.

serine or threonine residues. Such glycans are found on the protein dystroglycan, which forms part of a complex that connects the cytoskeleton to the extracellular matrix (Figure 13.6). The name dystroglycan reflects the role of this protein as the membrane anchor for the cytoplasmic actin-binding protein dystrophin, which in turn derives its name from the fact that the most common forms of muscular dystrophy result from changes in the sequence of this protein. In these cases, the muscle weakness characteristic of muscular dystrophy is believed to reflect a failure of dystrophin to maintain the proper architecture of muscle fibres. A total of more than 20 O-linked glycans are attached to a serine- and threonine-rich region in the middle of the α subunit of the dystroglycan polypeptide, forming an extended region between globular N- and C-terminal domains. The glycans are a mixture of typical GalNAc-linked mucin-type structures and unusual mannose-linked oligosaccharides. One form of mannose-linked glycan is a tetrasaccharide in which the mannose residue is extended with GlcNAc, galactose, and sialic acid residues. Glycosylated dystroglycan has a molecular mass roughly twice the 75 kDa mass of the core polypeptide. In addition to being expressed in muscle cells, dystroglycan with somewhat different glycosylation is found on the surface of neurones in the brain and eye.

Five congenital forms of muscular dystrophy result from amino acid changes in proteins which appear, based on sequence similarity, to be glycosyltransferases. These mutations result in distinct sets of symptoms involving various combinations of developmental defects in muscle, brain, and eye. Two of these proteins are now known to catalyse the first two steps (addition of mannose and GlcNAc) in the synthesis of O-linked mannose glycans. The mutations in these two proteins that cause muscular dystrophy result in a dramatic reduction in the glycosylation of α-dystroglycan, which reduces the molecular weight by about 60 kDa in cells from muscular dystrophy patients. Concomitant with the change in glycosylation, the ability of dystroglycan to bind to the extracellular matrix protein laminin is lost. This evidence indicates that mannose-linked O-glycans have an important role in the dystroglycan–laminin interaction, though the precise nature of this role is not yet known. The type G repeats in laminin bind to sugars, but inhibition studies with the terminal structures from simple mannose-linked tetrasaccharides and treatment of dystroglycan with glycosidases that degrade these glycans does not disrupt laminin binding to dystroglycan.

Some of the confusion about the role of mannose-linked O-glycans may be clarified as the activities of the other putative glycosyltransferases affected in muscular dystrophy are defined. At least one of these proteins, the product of the *LARGE* gene, catalyses the synthesis of novel glycans. This protein contains two potential glycosyltransferase catalytic domains, which have been suggested to have glucosyltransferase and GlcNAc transferase activity. By analogy to the bifunctional enzymes that synthesize glycosaminoglycan chains such as heparin, the LARGE protein might synthesize extended chains of repeating glucose-GlcNAc disaccharide units, for example. Such chains attached to the central domain of dystroglycan would account for the very large shifts in molecular weight when the glycans are removed and they would provide a target for laminin binding.

A major breakthrough in understanding the role of glycans on dystroglycan is the demonstration that the extended glycans synthesized by the LARGE protein are linked to the core mannose residue through a phosphodiester linkage (Figure 13.6). Mutations in POMGNT1 that prevent addition of GlcNAc to the O-linked mannose reduce the efficiency of phosphorylation, but the residual attachment of some phosphate allows elaboration of a reduced number of extended chains, which probably explains the milder symptoms associated with these mutations that cause Muscle–Eye–Brain disease. The presence of some phosphate may also explain why overexpression of LARGE can increase glycosylation of α-dystroglycan and restore laminin binding in muscle precursor cells from patients with mutations in POMGNT1, presumably by increasing the efficiency of addition of the elongated chains. However, the glycan made by the LARGE transferase can also be expressed on other N- and O-linked glycans when the normal acceptor is not available.

⊙ See section 5.3 for more on sequences of glycosyltransferases.

Because overexpression of LARGE can compensate for multiple genetic defects leading to muscular dystrophy, transgenic expression of LARGE might be a viable approach to treatment of the entire class of glycosylation-associated muscular dystrophies. There are other substrates for protein O-mannosyltransferase besides dystroglycan, with particularly high levels of the enzyme and of mannose-linked O-glycans found in brain. In epithelial tumour cells, reduction in the activity of the LARGE protein is associated with loss of adhesion, facilitating cancer progression, so *LARGE* can also be viewed as a tumour suppressor gene.

Overexpression of another glycosyltransferase has been suggested to have therapeutic potential for the most common forms of muscular dystrophy, which result from mutations in the dystrophin gene. The cytotoxic T cell or CT antigen is normally expressed at the neuromuscular junction. Overexpression of the CT GalNAc transferase, which is responsible for synthesis of this antigen, in dystrophic muscle prevents the muscle from degenerating. CT GalNAc transferase overexpression increases expression of utrophin, a dystrophin homologue, which replaces the defective dystrophin through a mechanism that is not understood.

13.7 Changes in glycosylation are associated with cancer

The most intensively studied correlations of disease phenotypes and changes in glycosylation are in the field of tumour biology. Many such changes have been identified using antibodies to carbohydrates, such as those that recognize glycans associated with specific blood groups or cell surface glycans normally seen only on cells in early stages of development. In the latter case, the structures recognized by the antibodies are referred to as oncofoetal antigens. Structural studies have confirmed that increases or decreases in reactivity with such antibodies in tumours compared with normal tissues result from specific changes in the carbohydrate structures present on secreted and cell surface glycoproteins. In many cases, these structural changes have been further correlated with changes in the activity of one or more glycosyltransferases during the process of transformation from a normal to a tumour cell. There are three principal reasons for the continuing interest in these changes:

- the possibility that changes in glycosylation can explain some of the phenotypic changes in tumour cells
- the use of specific changes in glycosylation as diagnostic markers of certain types of tumours
- the hope that the changes can be exploited in the treatment of cancer.

In the light of our current understanding of the role of genetic mutations and oncogenes in the generation of tumorigenic cells, the possibility that glycosylation changes represent the fundamental, initiating change in the transformation process can be dismissed. However, it is still important to know whether the changes in

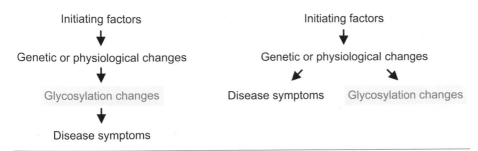

Figure 13.7 Possible relationships of glycosylation changes to development of symptoms in cancer and other diseases. In some instances, glycosylation changes may actually cause disease symptoms, so that interfering with the changes might be therapeutically useful. In diseases in which glycosylation changes are correlated with but not causally related to the disease symptoms, the changes may still be useful as diagnostic markers.

glycosylation form part of the chain of events linking genetic changes to uncontrolled growth of tumour cells and their ability to metastasize, or if they are simply bystander effects resulting from changes in the properties of the tumour cells (Figure 13.7). In the former case, correction of aberrant glycosylation might lead directly to therapeutic opportunities, but even in the latter, the glycosylation changes can be useful markers for detecting and attacking tumour cells (Box 13.1).

There are several ways in which glycosylation changes could lead to cancer-associated phenotypes in cells. For example, because glycosaminoglycans attached to cell surface proteoglycans play an important part in the action of growth factors at the surface of cells, it seems quite possible that changes in the structures of the glycosaminoglycans would affect the response of cells to growth factors and thus their capacity for replication. Similarly, the positions of glycans at cell surfaces and their roles in cell adhesion suggest that changes in glycosylation may be involved in the ability of metastatic cells to move from one site to another in the body. Glycosylation changes could result in loss of adhesion associated with cell dispersal from primary tumours, or they could initiate new cell adhesion processes that result in the establishment of secondary tumours at new sites. Because lectins such as the siglecs and particularly the selectins are known to mediate cell adhesion and extravasation of normal cells, there is certainly a precedent for a role for glycans and glycan-binding receptors in the types of adhesion events that must occur during metastasis and various models for their participation in metastasis have been proposed (Figure 13.8).

The metastatic potential of tumour cells has been extensively correlated with increases in the sialylation of cell surface glycoproteins. An important cause of this increased sialylation is increased branching of complex N-linked oligosaccharides in highly metastatic cells. Increased branching results from increased activity of GlcNAc-transferase V, which adds 1-6-linked GlcNAc to a typical bi-antennary structure to create a tri-antennary structure (Figure 13.9). The addition of terminal sialic acid to the extra branch helps to explain the overall increase in sialylation of

See **section 12.1** for more on proteoglycans and growth factors.

See **sections 9.5 and 9.6** for more on selectins.

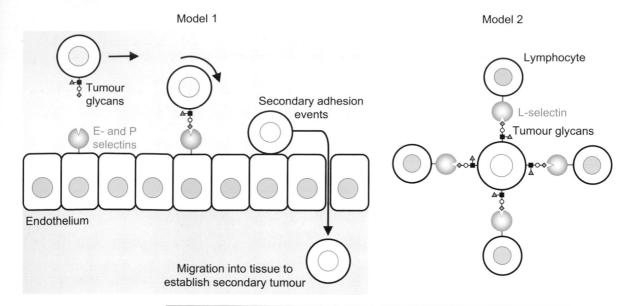

Figure 13.8 Potential roles of selectins binding to sialylated glycans in metastasis. Cells released from a primary tumour into the blood express sialyl-Lewis^X and related epitopes. In model 1, E- and P-selectin directly facilitate movement of tumour cells into tissues in the same way that they normally facilitate extravasation of neutrophils. In model 2, L-selectin binding to the glycans leads to lymphocytes adhering to tumour cells, forming micro-emboli that lodge in tissues.

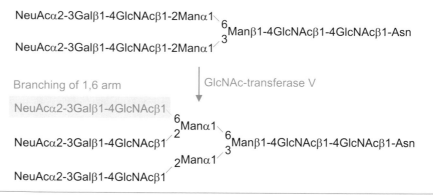

Figure 13.9 Increased branching of N-linked glycans resulting from activity of GlcNAc-transferase V in tumour cells. The addition of a further sialic acid-terminated branch may increase interactions of cell surface glycans with sialic acid-binding lectins.

these cell surfaces. In model systems, the activity of GlcNAc-transferase V in variant cells is well correlated with changes in metastatic potential, and selection of cells that display different levels of branching results in corresponding changes in metastatic behaviour. Inhibitors of N-linked glycan processing, such as swainsonine, reduce the level of sialylated, complex N-glycans on cell surfaces and these

Box 13.1   253

BOX 13.1   Glycotherapeutics: *Drugs that modify glycans on tumour cells may be useful in treating cancer*

The fact that some changes in glycosylation on the surface of tumour cells affect the progression of cancer may provide a basis for development of anti-cancer drugs. One approach is to modify the glycans expressed on tumour cells, as illustrated by recent attempts to modulate the expression of sialylated antigens on carcinomas. Carcinomas including colon, breast, and lung tumours derive from epithelial cells and are particularly prone to metastasis. Clinical and experimental evidence supports the idea that tumour cells have co-opted the selectin adhesion system in metastasis. For example, there is a good correlation between the degree of expression of sialyl-Lewis[x] and sialyl-Lewis[a] antigens on carcinomas and their metastatic potential. The fact that metastasis is diminished in mice lacking P-selectin provides further evidence that selectins play a role in metastasis

This information suggests that blocking of selectin–glycan interactions might provide a means of reducing metastasis. Direct blocking of the interaction has proven to be very difficult, because soluble ligands such as the sialyl-Lewis[x] tetrasaccharide are difficult to deliver as drugs and compete poorly for binding to cell surfaces. Such approaches were investigated in an attempt to control inflammation by inhibiting the selectins and they were not successful. Therefore, the alternative strategy of attempting to reduce the level of target glycan expression on tumour cells is being investigated. One approach is to inundate tumours with substrate decoy molecules that serve as primers and are extended by glycosyltransferases at the expense of their normal membrane glycoprotein substrates, thus reducing the glycosylation of cell surface glycoprotein. Because glycosyltransferases typically have high affinity for their acceptor substrates, the levels of decoys required to prevent selectin–ligand formation are much lower than the levels needed to inhibit the selectin–glycan interaction directly. In a related strategy, an acceptor lacking the 4-hydroxyl group is used directly as an inhibitor of galactosyltransferase. Appropriate affinities were achieved by using disaccharides. However, the drugs need to gain access to intracellular targets. Creating glycosides with large hydrophobic groups helps make the compounds less polar, but in order for them to penetrate the plasma membrane, they need to be delivered in per-O-acetylated form. Intracellular hydrolysis of the O-acetyl groups generates the free decoy and inhibitor molecules. Drug concentrations as low as 25 μM in the medium of cells result in intracellular levels of the free sugars that are sufficient to reduce the level of cell surface sialyl-Lewis[X] expression by 50%. Treated cells injected into mice show reduced ability to infiltrate organs and form colonies, thus providing an initial indication that the decoy strategy may be a viable therapeutic approach.

Essay topics

• Discuss the evidence that selectin–ligand interactions play an important role in tumour metastasis.

• Compare inhibitors of glycosylation and inhibitors of glycan–receptor interactions as potential therapeutic agents using cancer and influenza virus as examples.

Lead references

• Brown, J.R., Yang, F., Sinha, A., Ramakrishnan, B., Tor, Y., Qasba, P.K., and Esko, J.D. (2009). Deoxygenated disaccharide analogs as specific inhibitors of β1-4-galactosyltransferase 1 and selectin-mediated tumor metastasis, *Journal of Biological Chemistry* **284**, 4952–4959.

• Fuster, M.M., Brown, J.R., Wang, L., and Esko, J.D. (2003). A disaccharide precursor of sialyl Lewis x inhibits metastatic potential of tumor cells, *Cancer Research* **63**, 2775–2781.

• Kim, Y.J., Borsig, L., Varki, N.M., and Varki, A. (1998). P-selectin deficiency attenuates tumor growth and metastasis, *Proceedings of the National Academy of Sciences U.S.A.* **95**, 9325–9330.

• Weston, B.W., Hiller, K.M., Mayben, J.P., Manousos, G.A., Bendt, K.M., Liu, R., and Cusack, J.C., Jr. (1999). Expression of human α(1,3)fucosyltransferase antisense sequences inhibits selectin-mediated adhesion and liver metastasis of colon carcinoma cells, *Cancer Research* **59**, 2127–2135.

inhibitors can reverse some of the growth and metastatic properties of tumour cells *in vitro* and *in vivo*. Thus, it seems plausible that these specific glycosylation changes play a significant part in the behaviour of tumour cells. However, a specific molecular mechanism for the involvement of sialylation and branching changes in metastatic adhesion events has not been established.

13.8   Changes in glycosylation may be useful biomarkers for cancer detection and treatment

Even in the absence of a complete understanding of how changes in glycosylation relate to tumour progression, these changes can provide important avenues to cancer detection and treatment. Activation of some glycosyltransferases is particularly associated with certain types of tumours. Regardless of whether the resulting glycosylation changes are essential to the development or spread of such tumours, the changes can be useful diagnostic or prognostic indicators known as biomarkers. For example, the sialyl-Tn epitope, which consists of a truncated O-linked glycan NeuAcα2,6GalNAcα-Thr/Ser, is exposed at high levels on a variety of carcinomas. Monoclonal antibodies that recognize this structure with high affinity and specificity have been used to establish a correlation between expression and patient survival. The levels of various carbohydrate epitopes can also be used to monitor the effectiveness of chemotherapy.

Targeted killing of cancer cells by stimulating an immune response directed towards antigens specific to tumour cells has been a long-standing goal of cancer research. The biggest challenge in developing immunotherapy of this kind is to identify antigens that are truly unique to tumour cells so that tumour-specific antibodies can be developed. In passive immunotherapy, such antibodies can be conjugated to exogenous toxins, drugs, or isotopic labels that are then directed specifically to the tumours. Alternatively, immunogens based on tumour-specific glycans could be used to stimulate a cytotoxic response by the host immune system. The changes in glycosylation associated with transformation and the appearance of novel carbohydrate epitopes on tumour cells suggest that carbohydrates might serve as suitable targets in these strategies.

There are many changes in cell surface glycans brought on by oncogenic transformation, but these changes often result from relatively modest quantitative changes in glycosyltransferase activities in the luminal compartments of tumour cells. Thus, while the distribution of different glycan structures changes significantly, few structures are unique to the tumour cell surface. This situation makes identification of tumour-specific carbohydrate antigens quite difficult. Nevertheless, attempts to make carbohydrate vaccines continue, spurred by recent advances in oligosaccharide synthesis that can provide a homogeneous and essentially unlimited supply of antigens. Use of such vaccines has resulted in specific antibody responses, but it remains a significant challenge to direct an appropriate cytotoxic response. Many anticarbohydrate antibodies are of the IgM rather than IgG class, which can be less effective in targeting cytolysis. Effective responses may also require cytotoxic T cell responses, and development of antigens and adjuvants that evoke such responses may prove to be difficult.

⊖ See sections 6.10 and 6.11 for more on the chemical synthesis of glycans.

One potentially useful antigenic epitope associated with alterations in glycosylation in breast cancer is actually a peptide antigen. MUC-1 is a major cell surface mucin on breast epithelial cells. The peptide core of MUC-1 is an extended transmembrane protein, much of which consists of tandem repeated sequences of 20 amino acids that bear multiple sites of O-linked glycosylation. In normal breast epithelia, these sites are mostly occupied by core 2 structures, while in tumour cells the sites are predominantly linked to core 1 structures. This change is thus associated with a shift toward shorter O-glycan chains. The shift results from a change in the relative activity of the sialyltransferase, which terminates synthesis in the core 1 form, and the GlcNAc-transferase that forms core 2 (Figure 13.10). Reconstitution experiments demonstrate that these two enzymes compete for access to the substrate

⊖ See section 3.1 for more on mucins.

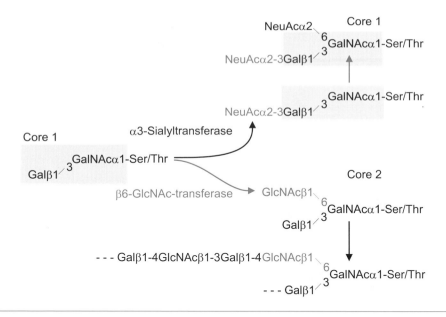

Figure 13.10 Competition between sialyl and GlcNAc-transferases in O-linked glycan synthesis. Addition of the β1-3-linked GlcNAc to core 1 creates core 2, which can be further elaborated. Addition of the α2-6-linked sialic acid prevents the action of the GlcNAc-transferase, so that the largest structure formed is the disialylated core 1 structure.

within the Golgi apparatus, so an increase in the sialyltransferase activity and a decrease in GlcNAc-transferase activity result in the observed shift in O-linked core structures. There is also evidence that fewer potential sites for the attachment of O-linked glycans may be utilized. As a consequence of these changes, portions of the MUC-1 core protein that are normally shielded from the immune system by carbohydrate become exposed in the tumour mucin. These exposed epitopes are tumour specific, making them good prospective targets for immunotherapy. In this case, the targeting epitope can be highly restricted to the tumour cells even though the changes in glycosyltransferase activities are modest.

Some widely used tests for biomarkers of cancer detect glycoproteins, for example the prostate-specific antigen test for prostate cancer and the carcinoembryonic antigen test for colon cancer, but these tests are not based on changes in glycosylation. The new technologies of glycoproteomics and glycomics are being applied to identify new biomarkers based specifically on glycan changes that may provide sensitive tests for early detection of cancer. One strategy is to profile serum for the presence of specific anticarbohydrate antibodies that may be generated against tumour-specific glycans which appear foreign to the immune system. Profiling can be done using glycan arrays to measure the binding of antibodies to libraries of glycans and glycopeptides with structures that are found in glycoproteins, such as MUC-1, that have known glycosylation changes associated with cancer. Using this technique, cancer-associated IgG antibodies to MUC-1 have been detected in serum from patients with

See **section 6.7** for more on glycan arrays.

ovarian, breast, and prostate cancer. The antibodies are against three different O-glycopeptide epitopes of MUC-1, containing sialyl-Tn, Tn, or truncated core 3 glycans. These preliminary findings are encouraging for the prospects of developing specific tests for cancer based on glycosylation changes.

SUMMARY

Demonstrating that changes in glycosylation have a direct role in the aetiology of disease continues to be a challenge. Some of the examples discussed in this chapter and in other parts of this book show that genetic diseases associated with mutations in glycan biosynthesis pathways and recognition systems can provide important insights into the processes mediated by glycans and their receptors. However, in many cases the changes brought about by such mutations can be subtle and unexpected. The analysis of the molecular basis for von Willebrand disease demonstrates how the role of glycosylation in the disease may only become evident when an appropriate genetic linkage is established. Large numbers of polymorphisms associated with disease and disease susceptibility are now being mapped. Our improving ability to identify genes that encode proteins that function in biosynthesis, transport, degradation, and recognition of glycoconjugates will undoubtedly facilitate the establishment of further links between glycans and disease processes. Knowledge of the way that changes in glycans and their receptors contribute to dysfunction at the level of cells, tissues, and organismal physiology may also point the way to effective therapeutic intervention.

KEY REFERENCES

Barresi, R. and Campbell, K.P. (2003). Dystroglycan: from biosynthesis to pathogenesis of human disease, *Journal of Cell Science* **119**, 199–207. This review and the paper by Schachter *et al.* (2004) (see below) summarize the genetics and biochemistry of dystroglycan glycosylation defects.

Brockhausen, I. (1999). Pathways of O-glycan biosynthesis in cancer cells, *Biochimica et Biophysica Acta* **1473**, 67–95. A detailed review of how O-glycan biosynthesis is altered in different types of cancer.

Bucala, R. and Cerami, A. (1992). Advanced glycosylation: chemistry, biology and implications for diabetes and aging, *Advances in Pharmacology* **23**, 1–34. Glycation is reviewed in detail.

Dennis, J.W., Granovsky, M., and Warren, C.E. (1999). Glycoprotein glycosylation and cancer progression, *Biochimica et Biophysica Acta* **1473**, 21–34. The evidence that increased branching of the 1–6 arm of N-linked glycans contributes to cancer progression is reviewed.

Dennis, J.W., Granovsky, M., and Warren, C.E. (1999). Protein glycosylation in development and disease, *BioEssays* **21**, 412–421. This paper provides a good overview of protein glycosylation in health and disease.

Freeze, H.H. and Westphal, V. (2001). Balancing N-linked glycosylation to avoid disease, *Biochimie* **83**, 791–799. This paper reviews congenital disorders of glycosylation and speculates on possible selective advantages of mutations in the N-linked glycosylation pathway.

Kim, Y.J. and Varki, A. (1997). Perspectives on the significance of altered glycosylation of glycoproteins in cancer, *Glycoconjugate Journal* **14**, 569–576. This review gives details of the different glycosylation changes seen in cancer and speculates on how these changes might be linked to cause or progression of tumours.

Macher, B.A. and Galili, U. (2008). The Galα1-3Galβ1-4GlcNAc-R (α-Gal) epitope: a carbohydrate of unique evolution and clinical relevance, *Biochimica et Biophysica Acta* **1780**, 75–88. The characteristics of the α-gal epitope, the antibodies that recognize it, and how this interaction causes xenograft rejection are reviewed.

Mohlke, K.L., Purkayastha, A.A., Westrick, R.J., Smith, P.L., Petryniak, B., Lowe, J.B., and Ginsburg, D. (1999). Mvwf, a dominant modifier of murine von Willebrand factor, results from altered lineage-specific expression of a glycosyltransferase, *Cell* **96**, 111–120. Experiments showing that misexpression of a GalNAc transferase causes von Willebrand factor deficiency are described.

Rich, J.R. and Withers, S.G. (2009). Emerging methods for the production of homogeneous human glycoproteins, *Nature Chemical Biology* **5**, 206–215. Some of the approaches being developed for glycosylation engineering in cells and glycan remodelling with enzymes are summarized.

Schachter, H. (2001). Congenital disorders involving defective N-glycosylation of proteins, *Cellular and Molecular Life Sciences* **58**, 1085–1104. These disorders are reviewed in detail.

Schachter, H., Vajsar, J., and Zhang, W. (2004). The role of defective glycosylation in congenital muscular dystrophy, *Glycoconjugate Journal* **20**, 291–300. See Barresi and Campbell (2003) above.

Tretter, V., Altmann, F., Kubelka, V., Marz, L., and Becker, W.M. (1993). Fucose α1,3 linked to the core region of glycoprotein N-glycans creates an important epitope for IgE from honeybee venom allergic individuals, *International Archives of Allergy and Immunology* **102**, 259–266. Experiments defining α1-3 linked fucose as a major cause of allergic responses to bee stings are presented.

Wandall, H.H., Blixt, O., Tarp, M.A., Pederson, J.W., Bennett, E.P., Mandel, U., Ragupathi, G., Livingston, P.O., Hollingsworth, M.A., Taylor-Papadimitriou, J., Burchell, J., and Clausen, H. (2010). Cancer biomarkers defined by autoantibody signatures to aberrant O-glycopeptide epitopes, *Cancer Research* **70**, 1306-1313. This paper describes production and screening of a glycan array of O-glycopeptides for identification of cancer-associated anti-carbohydrate antibodies.

Yoshida-Moriguchi, T., Yu, L., Stalnaker, S.H., Davis, S., Kunz, S., Madson, M., Oldstone, M.B.A., Schachter, H., Wells, L., and Campbell, K.P. (2010). *O*-Mannosyl phosphorylation of alpha-dystroglycan is required for laminin binding, *Science* **327**, 88–92. A surprising finding about the presence of a phosphodiester linkage in *O*-mannose-linked glycans of dystroglycan helps to explain characteristics of the congenital muscular dystrophies.

QUESTIONS

13.1 Describe different ways in which naturally occurring genetic mutations result in altered glycan synthesis or recognition. How have phenotypes arising from such mutations provided evidence for the functions of oligosaccharides?

13.2 Describe the types of experiments used to define the molecular basis for congenital disorders of glycosylation.

References: Kim, S., Westphal, V., Srikrishna G., Mehta, D.P., Peterson, S., Filiano, J., Karnes, P.S., Patterson, M.C., and Freeze, H.H. (2000). Dolichol phosphate mannose synthase (DPM1) mutations define congenital orders of glycosylation Ie (CDG-Ie), *Journal of Clinical Investigation* **105**, 191–198.

Burda, P., Borsig, L., de Rijk-van Andel, J., Wevers, R., Jaeken, J., Carchon, H., Berger, E.G., and Aebi, M. (1998). A novel carbohydrate-deficient syndrome characterized by a deficiency in glucosylation of the dolichol-linked oligosaccharide, *Journal of Clinical Investigation* **102**, 647–652.

13.3 Discuss the involvement of antibodies against carbohydrates in disease.

13.4 Assess the indirect evidence that the product of the *LARGE* gene is a glycosyltransferase that might synthesize chains of repeating glucose-GlcNAc disaccharide units.

Reference: Patnaik, S.K. and Stanley, P. (2005). Mouse LARGE can modify complex N- and mucin O-glycans on α-dystroglycan to induce laminin binding, *Journal of Biological Chemistry* **280**, 20851–20859.

13.5 Describe the experiments used to show that alterations in the glycan structures of MUC-1 on tumour cells are associated with changes in activity of glycosyltransferases.

Reference: Dalziel, M., Whitehouse, C., McFarlane, I., Brockhausen, I., Gschmeissner S., Schwientek, T., Clausen, H., Burchell, J.M., and Taylor-Papadimitriou, J. (2001). The relative activities of the C2GnT1 and ST3Gal-1 glycosyltransferases determine O-glycan structure and expression of a tumor-associated epitope on MUC1, *Journal of Biological Chemistry* **276**, 11007–11015.

14

The future of glycobiology

LEARNING OBJECTIVES

By the end of this chapter, students should understand:

1 The types of evidence needed to prove the importance of glycans in specific biological functions

2 The importance of extending glycomic studies to an ever smaller scale and the need for novel approaches to analysis of expression and glycomic data

3 The impact of genomics and new genetic approaches on our understanding of glycan synthesis and recognition

Previous chapters of this book have highlighted the functions that sugars can perform. Although additional paradigms may arise, the examples from glycoprotein structure and trafficking and cell adhesion and signalling probably illustrate the major roles that glycans have. However, while the general question 'What is the function of glycosylation?' has probably been answered satisfactorily, specific questions of the form 'What is the function of this particular glycan on this particular glycoconjugate?' will be the basis for a great deal of future work. This chapter highlights some of the ways in which this type of question can be addressed.

14.1 Biochemistry, cell biology, and genetics must be combined in order to define roles of glycans

The examples discussed in previous chapters represent only a fraction of the situations in which biological roles for carbohydrates and their receptors have been established. The cases chosen are those where we know the most about the structures of the relevant glycans and how they are recognized. In the most favourable cases, such as the selectins, we have genetic evidence from knockout mice that the proteins are needed for appropriate inflammatory responses as well as detailed biochemical understanding of how they interact with selected glycans and the type of cell–cell interactions they can mediate. This information is complemented by genetic analysis of the enzymes needed to make the appropriate carbohydrate ligands and structural characterization of the ligands on target cells. Together these studies convince us that we understand correctly the physiological roles of the selectins and their glycan ligands, even though further studies are needed to complete our understanding.

The selectin example is atypical, because of the detail of our understanding and the fact that it is supported by such a diversity of evidence. For most of the other examples discussed in this book, we have at least some complementary genetic and biochemical evidence, but in many other cases the connection between glycans and physiological processes has only been made through one of these approaches. Thus, the challenges for the future will be two-fold: to continue to use biochemical and genetic approaches to provide clues about possible functions of glycans and to bring together multiple lines of investigation to provide compelling evidence for the roles of individual glycan–receptor interactions (Figure 14.1).

14.2   Glycomics and systems glycobiology underpin our understanding of glycobiology

The glycans that are ultimately attached to specific glycoproteins in individual cells result from a complex interaction between the acceptor protein and the complement of glycosyltransferases expressed in the cell. In theory, a complete description of the pattern of expression of glycosyltransferases in a cell combined with a full understanding of the biochemical characteristics of each of the transferases would allow us

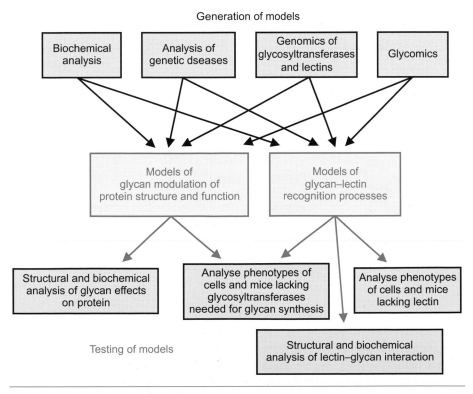

Figure 14.1 Sources of ideas about roles of glycans and lectins.

to predict what glycans will be synthesized. Computational methods of systems biology can be used with existing data to correlate the results of glycomic analysis of glycans with genomic analysis of glycosyltransferase expression to achieve some level of predictive capability. There is a continuing need to accumulate data that can be used to refine these predictive algorithms, particularly from comparable cell populations isolated from wild type and knockout mice lacking individual glycosyltransferases.

14.3 The cell- and protein-specific nature of glycosylation presents enormous technical challenges

Because glycosylation is a cellular event, correlations of glycan pools and glycosyltransferase expression would ideally be undertaken at the level of individual types of differentiated cells. Analytical methods that are needed for such studies will continue to be applicable to ever smaller, and hence increasingly homogeneous, populations of cells and the rate of throughput will continue to increase. Such studies of transcriptomes and glycomes will be most easily applied to cells that can be relatively easily isolated by methods such as cell sorting, and this means that cells of the immune system will be initial targets for this type of work. Alternative, more qualitative approaches will continue to be useful in solid tissues containing multiple cell types. In such cases, lectins and anticarbohydrate antibodies can be used to localize specific glycans and *in situ* hybridization with gene-specific probes can be used to identify cells expressing specific glycosyltransferases. These approaches will make it possible to track changes in glycosylation that occur during cellular differentiation and organismal development. Unique expression of specific glycans will provide important clues about novel roles of the glycans, which can then be investigated on a case-by-case basis.

One of the lessons from the selectin system is that target glycans involved in recognition processes can often be minor components of the total glycan pool, as illustrated by the fact that the essential sialyl-Lewis[X] structure is found only at a single site on P-selectin glycoprotein ligand 1. In these cases, the challenge is to identify the essential site and to analyse the relevant glycan. Such studies will continue to be performed using biochemical approaches, but they will benefit from increasingly sensitive analytical methods. The transient or dynamic nature of many forms of glycosylation also needs to be given attention. Some examples of glycosylation changes with biological consequences have been discussed in earlier chapters, but many others are only just coming to light.

14.4 Genomics provides critical insights into glycobiology

The complete sequences of the human and various animal genomes provide new approaches to the study of glycobiology. A key to learning about glycobiology from the genomic data is to identify genes encoding glycosyltransferases and lectins.

Three-dimensional structures of transferases and structure-based sequence comparisons have provided the information needed to identify novel transferases and it is likely that we know nearly the full complement of transferases in humans and other organisms. However, further analysis of the activities of expressed forms of the transferases is needed to define the role of each enzyme in glycan synthesis. Because the total number of transferases is in the hundreds rather than the thousands, the task is large but feasible. Knowing the specificities of the transferase families is an essential complement to the analysis of glycosylation in cells and organisms as described in the preceding sections.

Given the large number of terminal elaborations on glycans, one could envision a correspondingly large number of receptors that interact with different glycans. However, genomic analysis suggests that the actual number of lectin–glycan interactions is modest. Sequence motifs used to identify families of carbohydrate-recognition domains often identify lectin-like domains that have similar folds but lack sugar-binding activity, so knowledge of how carbohydrate-recognition domains interact with glycans is needed to assess the number of true sugar-binding proteins. Sugar-binding activity has evolved repeatedly in different structural contexts, so lectins that mediate important biological processes might not be identified in a screen of the genome based on the structures of known lectin families. Instances in which genetic evidence points to a critical role for glycan recognition, but where no relevant receptor has been identified, suggest that there may be categories of sugar-binding receptors remaining to be described. Novel strategies will be needed to identify such lectins. Particularly in the context of development, where only a very small number of cells may be involved in a specific interaction, genetic analysis may be critical to the identification of potential receptors which can then be studied biochemically.

As more lectins are identified, it will be essential to define their recognition properties and the cellular functions such as trafficking, adhesion, and signalling that they mediate. Glycan arrays have emerged as a powerful tool for characterizing the properties of glycan-binding receptors. Some further technical advances, such as presentation of glycans in a membrane-like format that allows lateral mobility, need to be investigated. However, a major task at the moment is to make the complement of glycans on the arrays as comprehensive as possible and to extend the range from mammalian glycans to include sugars from bacteria and fungi in order to facilitate characterization of receptor binding to pathogens as well as to target glycans on mammalian cells.

14.5   Traditional genetics will be combined with new approaches to human genetics to understand glycans and their receptors

Genetic studies, including random mutagenesis, targeted mutagenesis, and analysis of polymorphisms will continue to be a significant way to learn about the roles of glycans and their receptors. As demonstrated by a number of the examples discussed in previous chapters, analysis of the phenotypes of knockout mice lacking specific types of glycans or lectins can provide enormous insights into the physiological functions of glycan–receptor interactions. This approach will certainly continue to be one of the main routes to understanding the roles of glycans *in vivo*. Many of the knockout mice relevant to glycobiology have been produced and plans are in place to produce comprehensive panels of knockouts covering the mouse genome. The major challenge is going to be to analyse the phenotypes of these mice to understand what they tell us about how glycans function.

Mapping of mutations that are associated with human diseases also continues to be an important source of information about glycan functions. In addition to the examples discussed in earlier chapters, other diseases have been linked to glycosylation in this way, although in many cases the normal functions and the mechanisms of dysfunction remain to be established. When such mutations map to genes that encode glycosyltransferases, nucleotide sugar transporters or lectins, models can be constructed for how these proteins may function in normal human biology and how the mutations might cause disease. The availability of a large and growing body of information on genetic variations in humans provides an important additional way to correlate more subtle alterations in biochemical function with changes in physiology that are sometimes related to susceptibility to disease. Analysis of genes for sugar-binding receptors suggests that polymorphisms that affect the activities of these proteins are relatively common. In cases where there are no mouse orthologues of human biosynthetic enzymes and glycan-binding receptors, the human genetic data will be essential to establishing the physiological roles of species-specific glycosylation.

Random mutagenesis followed by selection of specific phenotypes will also continue to yield insights into glycan function. In general, such studies will be conducted with simpler model organisms. Glycans play structural roles in all of the evolutionary kingdoms, and their roles in protein folding and intracellular trafficking appeared early in the evolution of eukaryotes. Thus, lectins in the secretory pathway can be studied in yeast, but glycans and lectins that act as cell adhesion receptors must be studied in more complex organisms. For developmental studies, organisms such as zebrafish, *Drosophila* and *Caenorhabditis elegans* have clear advantages. However, it will be important to bear in mind that, within the animal kingdom, invertebrate and vertebrate glycans and lectins differ substantially. Thus, model organisms will continue to provide paradigms for the types of functions that glycans can perform, but the roles of individual glycans will sometimes be specific to a particular organism.

14.6   Molecular understanding of how glycans function will require further elucidation of structure–function relationships

A major ongoing aim of glycobiology is to describe how the structures and physical properties of glycans make them uniquely suited to perform certain functions.

Structural studies of glycan–receptor interactions will form the basis for asking the broader question of why some cellular recognition processes involve protein–carbohydrate interactions while many more involve protein–protein interactions. Some of the best characterized glycan–lectin complexes are involved in dynamic interactions, such as selectin-mediated leucocyte rolling. In such cases, physical characteristics of the protein–sugar interactions may be specifically advantageous. More generally, it may be possible to learn how the enzymes that synthesize glycans have co-evolved in a special relationship with the lectins that recognize these glycans.

14.7   Our increasing knowledge about glycobiology is being applied to practical issues

Applications of glycobiology in biotechnology follow from the knowledge of glycan structures and functions as described in several of the boxes highlighting glyco-therapeutics. The future is likely to see increased development of small molecules that are inhibitors of specific steps in glycan synthesis and processing as a way to modulate the activities of glycans in disease. In addition, rapid progress is being made in new technologies for producing recombinant therapeutic glycoproteins, as the field enters a new phase of engineering yeast and bacterial expression systems. Increased knowledge of receptor-mediated glycoprotein trafficking is also likely to suggest novel ways to use specific terminal sugar residues to target glycoproteins to specific tissues that express appropriate receptors.

SUMMARY

A great deal has been learned about the roles of glycans, but there are also many indications that glycans have functions that we do not yet understand. A range of approaches, from glycan and lectin structure analysis to physiological study of knockout mice, will be needed to provide convincing evidence for these roles. As our understanding of glycobiology broadens to encompass more examples, new routes to the application of this knowledge in the diagnosis and treatment of disease will undoubtedly emerge.

ABO blood group substances: Carbohydrate structures found attached to membrane and secreted proteins that determine the blood group of an individual. Individuals may contain the A, B, or both A and B antigenic substances, or else lack these substances (type O).

Adaptive immunity: Also known as acquired immunity, this is the immune response that develops in response to stimulation by specific antigens.

Adhesin: A protein found on the surface of a microorganism that binds to ligands on the surface of a host cell.

Advanced glycation end products (AGEs): Products resulting from reaction of high concentrations of glucose with amino acid side chains in proteins.

Aggrecan: A chondroitin sulphate/keratan sulphate proteoglycan found mainly in cartilage. It contains more than 100 glycosaminoglycan chains attached to the protein core.

Anomer: One of a pair of stereoisomers that differ only in the configuration about the anomeric carbon atom.

Anomeric carbon: The anomeric carbon is the new chiral centre created in forming the cyclic structure of a monosaccharide.

Anomeric configuration: The configuration about the anomeric carbon. Either α or β.

Asialoglycoprotein receptor: A galactose-binding C-type lectin involved in clearance of circulating glycoproteins by the liver. The receptor, found on hepatocytes, binds glycoconjugates with terminal galactose or GalNAc residues, and mediates their endocytosis.

Asn-X-Ser or Asn-X-Thr: The sequence in a protein required for attachment of N-linked oligosaccharides. The oligosaccharides are attached to the amide nitrogen of the asparagine residue. X can be any amino acid except proline.

Bi-antennary: Having two antennae. Refers to a type of N-linked complex oligosaccharide that is formed by addition of sugars to both the 1–6 arm and the 1–3 arm of the trimannose core.

Biomarker: An assayable change in the levels of a biological compound that can be correlated with the presence or progression of a disease.

Calnexin: A membrane-bound protein of the endoplasmic reticulum. Its luminal domain binds to the terminal glucose residue of the $Glc_3Man_9GlcNAc_2$ oligosaccharide attached to newly synthesized glycoproteins, retaining the glycoproteins in the endoplasmic reticulum while they fold.

Calreticulin: A soluble protein, homologous to calnexin, found in the lumen of the endoplasmic reticulum. Like calnexin, it binds to the terminal glucose residue of the $Glc_3Man_9GlcNAc_2$ oligosaccharide attached to newly synthesized glycoproteins and takes part in a quality control process that prevents incorrectly folded proteins from leaving the endoplasmic reticulum.

Carbohydrate-recognition domain (CRD): The part of a glycan-binding protein (lectin) that contains the sugar-binding site. Lectins are classified into different families based on similarities in sequence of their carbohydrate-recognition domains.

Cation-dependent mannose 6-phosphate receptor (CD-MPR): The smaller of two structurally related receptors needed to target lysosomal enzymes from the Golgi to the lysosomes. CD-MPR requires divalent cations to bind to mannose 6-phosphate residues on the glycans attached to the enzymes.

Cation-independent mannose 6-phosphate receptor (CI-MPR): The larger of two structurally related receptors needed to target lysosomal enzymes from the Golgi to the lysosomes. CI-MPR does not require divalent cations to bind to mannose 6-phosphate residues on the glycans attached to the enzymes.

CD2: Also known as LFA-2. A transmembrane protein of the immunoglobulin superfamily found on the surface of T cells. CD2 functions as a cell adhesion molecule, mediating adhesion between T cells and target cells as part of the immune response.

CD22: A sialic acid-binding protein found on B lymphocytes. One of the siglec family of I-type lectins.

CD34: A transmembrane glycoprotein expressed on endothelial cells and on hematopoietic stem cells that acts as a ligand for L-selectin during lymphocyte adhesion. CD34 is highly glycosylated with O-linked glycans attached to mucin-like domains

Ceramide: The simplest type of sphingolipid consisting of a fatty acid chain attached through an amide linkage to the amino alcohol sphingosine. Ceramide is the lipid constituent of glycosphingolipids, which have one or more sugar residues attached in β linkage through the 1-hydroxyl of ceramide.

Chondroitin sulphate: A glycosaminoglycan found on proteoglycans in cartilage, bone, and skin. Composed of repeating disaccharide units of glucuronic acid and *N*-acetylgalactosamine.

Chorionic gonadotropin: A hormone produced by the placenta. It stimulates the ovary to produce progesterone which is critical for maintenance of pregnancy. Chorionic gonadotropin is structurally related to follitropin and lutropin, having an identical α subunit but a unique β subunit.

Collectin receptor: A receptor that binds to the collagenous region of collectin molecules, and mediates internalization and destruction of the collectin with its bound ligand.

Collectins: A group of soluble C-type lectins composed of C-type CRDs attached to collagen-like domains. Examples are serum mannose-binding protein and lung surfactant proteins A and D.

Collectins initiate innate immune responses against micro-organisms.

Complement pathway: A cascade of serum proteins that leads to killing of a micro-organism as part of the immune response. The classical pathway is initiated by binding of the protein C1q to an antibody bound to a pathogen, whereas the lectin branch of the complement pathway is initiated by binding of mannose-binding protein to carbohydrates on the surface of the pathogen.

Complex oligosaccharide/glycan: An N-linked oligosaccharide that results from addition of sugars such as GlcNAc, galactose, and sialic acid to the trimannose core in the Golgi.

Concanavalin A (ConA): A plant (legume) lectin isolated from jack beans. It exists as a tetramer of four identical subunits each containing a sugar-binding site specific for mannose and glucose.

Congenital disorders of glycosylation (CDGs): A group of diseases caused by defects in enzymes or transporter proteins involved in oligosaccharide biosynthesis, resulting in underglycosylation of glycoproteins.

C-type CRD: The calcium-dependent sugar-binding domain characteristic of a C-type lectin.

C-type lectin-like domain: A domain homologous to the C-type CRDs of C-type lectins which shares the same fold but is missing some or all of the residues that ligate sugar and/or calcium.

C-type lectins: A group of glycan-binding proteins (lectins), found in animals, that bind sugars in a calcium-dependent manner. They are characterized by carbohydrate-recognition domains (CRDs) with a conserved series of amino acid residues, some of which form the overall fold while others bind to sugar and calcium. Examples are the selectin cell adhesion molecules, serum mannose-binding protein and the asialoglycoprotein receptor.

DC-SIGN (dendritic cell-specific ICAM-3-grabbing non-integrin): A membrane-bound mannose-binding C-type lectin found on dendritic cells. It is involved in mediating interactions between dendritic cells and T lymphocytes, as well as between dendritic cells and pathogenic micro-organisms.

DC-SIGNR (DC-SIGN-related protein): Also called L-SIGN. A membrane-bound mannose-binding C-type lectin similar in sequence and structure to DC-SIGN. Found on endothelial cells.

Decay-accelerating factor (DAF, CD55): A glycosylphosphatidylinositol-anchored membrane protein that protects self cells from attack by complement.

Dermatan sulphate: A specific type of chondroitin sulphate glycosaminoglycan found on proteoglycans in skin and blood vessels. Composed of repeating disaccharide units of iduronic acid or glucuronic acid and N-acetylgalactosamine.

Diacylglycerol: Glycerol with long chain fatty acids substituted at hydroxyl groups 1 and 2. An intermediate in the biosynthesis of phosphatidyl phospholipids.

Dolichol: An isoprenoid lipid with 15–19 isoprenoid units and a terminal phosphorylated hydroxyl group. Dolichol acts as a membrane bound-carrier for sugars in the synthesis of glycoproteins and glycolipids.

Enantiomer: One of two mirror image forms of a molecule with one or more chiral centres.

Endoglycan: A mucin-like transmembrane glycoprotein expressed on endothelial cells and on hematopoietic stem cells that acts as a ligand for L-selectin during lymphocyte adhesion. Structurally related to CD34.

Endoglycosidase: An enzyme that can hydrolyse an inner glycosidic linkage of an oligosaccharide or polysaccharide. For example, Endoglycosidase H can cleave between the two GlcNAc residues in the core of high-mannose oligosaccharides and some hybrid oligosaccharides.

Endoplasmic reticulum-associated protein degradation (ERAD): The process in which misfolded proteins are translocated back across the endoplasmic reticulum membrane and targeted for degradation by the proteasome.

Endoplasmic-reticulum-Golgi intermediate compartment (ERGIC): A dynamic, membranous compartment involved in transport of proteins from the endoplasmic reticulum.

Epidermal growth factor receptor: A cell surface receptor that binds epidermal growth factor.

The receptor is a dimer with cytoplasmic regions containing tyrosine kinase domains. Binding of epidermal growth factor to the extracellular region of the receptor causes activation of the intracellular tyrosine kinase domains, triggering signalling cascades that stimulate cell growth.

Epimer: A compound such as a sugar that differs in configuration about a single asymmetric centre.

Epimerization: The process in which the configuration about one chiral centre of a compound is inverted to give the opposite configuration.

ERGIC-53: A mannose-binding L-type lectin involved in transport of glycoproteins between the endoplasmic reticulum and the endoplasmic-reticulum-Golgi intermediate compartment (ERGIC).

Erythropoietin: A hormone, produced by the kidney that stimulates production of red blood cells.

E-selectin: A selectin cell adhesion molecule expressed by endothelial cells. E-selectin mediates adhesion of leucocytes to the endothelium at sites of inflammation.

E-selectin ligand 1 (ESL-1): A transmembrane glycoprotein of leucocytes that is bound by E-selectin during adhesion of leucocytes to the endothelium at sites of inflammation. ESL-1 has N-linked oligosaccharides with terminal sialyl-Lewis^x groups that are recognized by E-selectin.

Exoanomeric effect: An electronic phenomenon that affects the preferred conformation of sugars.

Exocyclic: External to a chemical ring structure.

Exoglycosidase: An enzyme that releases a monosaccharide from the non-reducing (outer) end of an oligosaccharide or polysaccharide by hydrolysis of the glycosidic linkage.

Fabry disease: A glycolipid storage disease caused by deficiency of the enzyme α-galactosidase A that catalyses degradation of globotriaosylceramide.

Fibroblast growth factor: One of a family of proteins that stimulate growth of a wide variety of cell types. Fibroblast growth factors bind heparan sulphate chains of proteoglycans which act as co-receptors for the growth factors during signalling.

Follitropin: Also known as follicle stimulating hormone (FSH). A hormone produced in the pituitary. Follitropin stimulates maturation of the follicles in the ovary and production of oestrogen. Follitropin is structurally related to lutropin and chorionic gonadotropin, having an identical α subunit but a unique β subunit

Functional glycomics: Investigation of the biological roles of glycosylation, combining analysis of glycan structure and recognition to elucidate the functions of glycans in cells and organisms.

Galactosphingolipid: A class of glycosphingolipids that have galactose as the first sugar attached to ceramide.

Galactosyltransferase: An enzyme that transfers galactose from the activated donor GDP-galactose to an acceptor oligosaccharide during synthesis of glycoproteins or glycolipids.

Galectins: A group of soluble animal lectins that bind ligands containing β-linked galactose. Their functions include modulation of cell adhesion and cell signalling through cross-linking of cell surface glycans.

Ganglioside: A type of glycosphingolipid originally isolated from brain. Gangliosides have at least three sugar residues attached to ceramide with the outer residue being sialic acid.

Gaucher disease: A glycolipid storage disease caused by deficiency of the enzyme β-glucocerebrosidase that catalyses degradation of glucosylceramide.

GlcNAc-transferase I: An enzyme of the medial Golgi that transfers GlcNAc from UDP-GlcNAc to the mannose residue on the 1–3 branch of the core oligosaccharide containing five mannose residues in the N-linked glycosylation pathway. This reaction is the first step in the production of complex or hybrid oligosaccharides.

Globoside: A glycosphingolipid based on the core structure GalNAcβ1-3Galα1-4Galβ1-4Glc.

Glucopyranose: The cyclic form of glucose containing a pyranose ring (a ring of five carbon atoms and one oxygen atom).

Glucosidase I: The enzyme that removes the terminal α1-2-linked glucose residue from Glc$_3$Man$_9$GlcNAc$_2$ during N-linked oligosaccharide biosynthesis.

Glucosidase II: The enzyme that acts after glucosidase I to remove the two α1-3-linked glucose residues during N-linked oligosaccharide biosynthesis.

Glucosphingolipid: A class of glycosphingolipids that have glucose as the first sugar attached to ceramide.

Glycan: A general term for the sugars found in glycoproteins or glycolipids.

Glycan array: A panel of immobilized glycans that can be probed with glycan-binding proteins to determine their binding specificity.

Glycation: Bonding of a reducing sugar such as glucose to proteins in a non-enzymatic reaction. It occurs in uncontrolled diabetes when blood glucose concentration is very high.

Glycocalyx: The network of glycans that project from the cell surface.

Glycoconjugate: A protein or lipid with sugars attached.

Glycoform: One of the differentially glycosylated forms of a glycoprotein. Glycoforms of a glycoprotein have the same protein sequence but differ in the number and/or structure of oligosaccharides attached.

Glycolipid: A lipid with one or more sugar residues attached.

Glycome: The total complement of glycans found in a cell, tissue or organism.

Glycophorin A: The main transmembrane protein found on the surface of red blood cells. Its extracellular region contains many sites for attachment of O-linked glycans.

Glycophospholipid: A glycolipid formed by attachment of one or more sugars to a phosphatidylglycerol core.

Glycoprotein: A protein containing covalently linked sugar molecules that are added as co-translational or post-translational modifications.

Glycosaminoglycan: A long unbranched polysaccharide molecule made up of repeating disaccharide units each containing one hexose derivative and one amino sugar-derivative.

They are negatively charged due to the presence of sulphate and uronic acid groups. Glycosaminoglycans form the glycan part of proteoglycans.

Glycosidase: An enzyme that hydrolyses a glycosidic linkage.

Glycosidic linkage: The bond linking monosaccharides in disaccharides, oligosaccharides, and polysaccharides. Formed by a condensation reaction between two hydroxyl groups, one from each of the two monosaccharides.

Glycosphingolipid: A glycolipid formed by attachment of one or more sugar residues to ceramide.

Glycosylated cell adhesion molecule 1 (GlyCAM-1): A mucin-like protein containing multiple O-linked glycans, secreted by endothelial cells. It associates with the plasma membrane of high endothelial venules of peripheral lymph nodes and acts as a ligand for L-selectin during lymphocyte homing.

Glycosylation engineering: The process of generating glycoproteins with altered, defined glycosylation either during biosynthesis or by remodelling *in vitro*.

Glycosylphosphatidylinositol (GPI) anchor: A glycolipid molecule that attaches some proteins to the plasma membrane. GPI anchors are composed of a phosphatidylinositol group linked through several carbohydrate residues to phosphoryl ethanolamine that attaches to the C-terminus of the protein. The two fatty acid chains of the phosphatidylinositol group anchor the protein to the membrane.

Glycosyltransferase: An enzyme that catalyses the transfer of a sugar from a donor molecule to an acceptor molecule to form a glycosidic link.

Glypicans: A family of heparan sulphate proteoglycans, attached to the cell surface by glycosylphosphatidylinositol anchors, which control growth factor signalling in many tissues.

Hexose: A monosaccharide with six carbon atoms.

High-mannose oligosaccharide/glycan: An N-linked oligosaccharide consisting of 5–9 mannose residues attached to two N-acetylglucosamine residues. Formed by removal of glucose and mannose residues from the $Glc_3Man_9GlcNAc_2$ precursor oligosaccharide added to proteins in the N-linked glycosylation pathway.

Hyaluronic acid: A polymer of repeating disaccharide units of glucuronic acid and N-acetylglucosamine found in the extracellular matrix. It can be of very high molecular weight.

I-cell disease: A lysosomal storage disorder also called mucolipidosis II which is characterized by a lack of multiple lysosomal enzymes in the lysosomes. The disease is caused by a deficiency of the N-acetylglucosamine-1-phosphotransferase enzyme needed to add mannose 6-phosphate to lysosomal enzymes in order for them to be targeted to lysosomes.

Immunodeficiency: A disorder in which parts of the immune system do not function properly, resulting in susceptibility to infection.

Innate immunity: Immunity that is naturally present, not acquired in response to a specific antigen. It is the first line of defence against infection and involves receptors that recognize a broad range of pathogens.

Insulin-like growth factor II: A growth factor related in sequence to proinsulin. It is expressed mainly in embryonic and foetal tissues, and regulates foetal growth.

Insulin receptor: A cell surface receptor that binds insulin. It is similar in overall organization and function to the epidermal growth factor receptor, but signalling downstream from the receptor stimulates energy storage and reduces energy mobilisation.

I-type CRD: A carbohydrate-recognition domain that is homologous to an immunoglobulin-like domain.

I-type lectins: Glycan-binding proteins (lectins) containing carbohydrate-recognition domains homologous to immunoglobulin-like domains (I type CRDs). Examples include sialoadhesin on macrophages, CD22 on B lymphocytes, and other siglecs which contain immunoglobulin-like domains that bind to sialic acid.

Keratan sulphate: A glycosaminoglycan found on proteoglycans in cartilage, cornea, and vertebral discs. Composed of repeating disaccharide units of galactose and N-acetylglucosamine.

Kin recognition: Association of glycosyltransferases that catalyse related steps in glycosylation.

α-Lactalbumin: A modifier protein that is one of the subunits of lactose synthetase.

Lactose synthetase: A soluble milk protein, consisting of galactosyltransferase and α-lactalbumin, which catalyses synthesis of the disaccharide lactose.

Lectin: A glycan-binding protein. Enzymes with sugar substrates and sugar-binding antibodies are usually not classed as lectins.

L-selectin: A selectin cell adhesion molecule found on lymphocytes.

L-type lectins: Glycan-binding proteins (lectins) with carbohydrate-recognition domains homologous to the sugar-binding regions of leguminous plant lectins. Examples are ERGIC-53 and VIP36.

Lutropin: Also called luteinizing hormone (LH). A hormone produced in the pituitary. Lutropin stimulates ovulation. Lutropin is structurally related to follitropin and chorionic gonadotropin, having an identical α subunit but a unique β subunit.

Leucocyte adhesion deficiency type II (LAD-II): A rare disease caused by a defect in the transporter that carries GDP-fucose, the donor sugar for fucosyltransferase reactions, into the Golgi from the cytoplasm. Fucose-containing oligosaccharides, including ligands for the selectins needed for leucocyte adhesion, cannot be synthesized.

Lysin motif (LysM): A sequence motif that defines a family of domains, broadly distributed in prokaryotes and eukaryotes, that are often involved in peptidoglycan-binding.

Lysosomal storage disorder: An inherited disease caused by deficiency of any of the lysosomal enzymes. Symptoms result from accumulation in the lysosomes of the macromolecule, for example a glycolipid or a polysaccharide that should be broken down by the deficient enzyme.

Major histocompatibility complex (MHC): A group of genes, on chromosome 6 in humans, that codes for many proteins with roles in the immune system. It was initially identified as being the region coding for cell surface proteins that determine whether transplants will be rejected, hence the term histocompatibility.

Mannose 6-phosphate receptors: Two receptors, the cation-dependent mannose 6-phosphate receptor and the cation-independent mannose 6-phosphate receptor, that target newly synthesized lysosomal enzymes from the Golgi to the lysosomes by binding to mannose 6-phosphate residues on the oligosaccharides of the enzymes.

Mannose-binding protein (MBP): Also known as mannose-binding lectin (MBL). A mannose-binding C-type lectin, found in serum, that plays a part in the innate immune response against pathogens. One of the collectins, it mediates fixation of complement through interaction of a collagenous domain with serine-proteases (MASPS) following binding via C-type CRDs to sugar residues on the surfaces of microorganisms.

Mannose receptor: An endocytic C-type lectin found on macrophages and liver endothelial cells. It binds and mediates endocytosis of glycoconjugates with terminal mannose, N-acetylglucosamine, or fucose.

Mannosidase: An enzyme that catalyses release of a mannose residue from an oligosaccharide.

MBP-associated serine proteases (MASPs): Serine-proteases that bind to mannose-binding protein (MBP) and become activated when MBP binds to carbohydrate ligands on a micro-organism. Activated MASPs initiate activation of the complement cascade, leading to killing of the micro-organism, by activating complement components C4 and C2.

M-type lectins: Glycan-binding proteins (lectins) with carbohydrate-recognition domains homologous to processing mannosidase enzymes. Examples include EDEM, which is involved in the endoplasmic reticulum-associated protein degradation pathway.

MUC-1: A transmembrane mucin expressed mainly on epithelial cells. Aberrantly glycosylated forms are expressed on human epithelial tumour cells such as those found in breast or ovarian cancer.

MUC-2: The main secretory mucin found coating the epithelia of the digestive tract, airways, and other organs with mucous membranes.

Mucins: Very large, highly glycosylated proteins with many O-linked glycans. Mucin polypeptides contain repeated regions of about 100 amino acids rich in serine and threonine residues to which the glycans are attached. Some mucins are membrane bound, but most are secreted at mucosal surfaces.

Mucin-like domain: A region of a protein containing multiple serine or threonine residues with mucin type O-linked glycans attached.

Mucosal addressin cell adhesion molecule 1 (MadCAM-1): A mucin-like transmembrane protein found on mucosal surfaces. It acts as a ligand for L-selectin during targeting of leucocytes into mucosal and inflamed tissues.

Myelin-associated glycoprotein (MAG): A sialic acid-binding protein found at the surface of myelin-forming Schwann cells and oligodendrocytes in the nervous system. One of the siglec group of I-type lectins.

Myelin sheath: The lipid-rich layer that coats the axons of nerve cells. It acts as an electrical insulator to allow efficient conduction of nerve impulses.

Neoglycoprotein: A synthetic glycoconjugate formed by chemical attachment of monosaccharides or oligosaccharides to a protein that is not naturally glycosylated.

Neolactoside: A glycosphingolipid based on the core structure Galβ1-4GlcNAcβ1-3Galβ1-4Glc.

Neural cell adhesion molecule (NCAM): A transmembrane glycoprotein of the immunoglobulin superfamily expressed on neural cells. It is involved in both adhesive and anti-adhesive interactions between cells.

Neuraminidase: Commonly used as an alternative name for sialidase. Strictly speaking, neuraminidases are enzymes that cleave unmodified neuraminic acid, but the name is often used to refer to enzymes that release modified forms such as NeuAc and other sialic acids.

Neurocan: A chondroitin sulphate proteoglycan found in nervous tissue.

N-linked oligosaccharide/glycan: Linked through a nitrogen atom. Refers to glycans attached to proteins through a bond to the amide nitrogen of an asparagine residue.

Nod factors: N-acylated oligomers of β1-4-linked N-acetylglucosamine produced by nitrogen fixing bacteria. They direct formation of root nodules on the roots of leguminous plants that are colonized by the bacteria.

Nodes of Ranvier: Regularly spaced gaps in the myelin sheath around axons that expose the membrane and allow electrical excitation to occur.

Non-reducing end: The portion of a glycan in which the carbonyl groups at C1 of the constituent monosaccharides (C2 in neuraminic acid) are sequestered in glycosidic bonds.

Notch receptor: A signalling receptor involved in developmental tissue patterning. It interacts with two main protein ligands, Delta and Serrate. The ligand specificity of Notch is modulated by modification of O-linked glycans attached to epidermal growth factor-like domains.

Nuclear Overhauser effect: A phenomenon observed in NMR spectroscopy due to transfer of spin polarization between atomic nuclei. It can be used to determine the distances between atoms not connected by covalent bonds via measurement of space interactions.

Nucleotide sugar donor: The activated sugar molecule used by a glycosyltransferase enzyme in oligosaccharide biosynthesis. The activated sugar consists of a nucleotide attached to a monosaccharide. Examples are UDP-GlcNAc and CMP-sialic acid.

Oligodendrocyte: A cell found in the brain and spinal cord that produces the myelin sheath for coating the axons of neurons.

Oligogalacturonides: Oligomers of α1,4-linked galacturonic acid released from plant cell walls as part of the response to infection or wounding of the plant.

Oligosaccharins: Oligosaccharides that function as signalling molecules in plants.

Oligosaccharyltransferase: The enzyme that transfers the precursor oligosaccharide from dolichol to nascent polypeptide chains in the N-linked glycosylation pathway. Oligosaccharyltransferase is a multi-subunit complex associated with the endoplasmic reticulum membrane.

O-linked oligosaccharide/glycan: Linked through an oxygen atom. Refers to glycans attached to proteins through a bond to the hydroxyl group of a serine or threonine residue.

Oncofoetal antigens: Antigenic proteins or glycans that are normally expressed during foetal development but which also undergo changes in expression levels on some tumours.

OS-9: A protein of the endoplasmic reticulum which has a P-type CRD and binds high mannose oligosaccharides on misfold glycoproteins prior to their export into the cytoplasm.

Pentose: A monosaccharide with five carbon atoms.

Peptide N-glycanase (PNGase): An enzyme that cleaves between the innermost N-acetylglucosamine and asparagine residues of N-linked oligosaccharides. It can be used to release N-linked glycans from glycoproteins for analysis.

Perlican: A cell surface heparan sulphate proteoglycan.

Permethylation: Reaction of all the free hydroxyl groups of a glycan with methyl groups.

Pili: Also called fimbriae. Thin, protein tubes extending from the cytoplasmic membrane of bacteria. Each pilus has a shaft made of a protein called pilin, and an adhesin at the tip that binds glycans on a host cell.

Proteoglycan: A specialized class of heavily glycosylated glycoprotein found in the extracellular matrix and at cell surfaces. They consist of a core protein with one or more glycosaminoglycan chains attached.

P-selectin: A selectin cell adhesion molecule found on endothelial cells and platelets. P-selectin mediates adhesion of leucocytes and platelets to the endothelium at sites of inflammation.

P-selectin glycoprotein ligand 1 (PSGL-1): The main glycoprotein ligand bound by P-selectin. It is a transmembrane protein, found on leucocytes, with an extended mucin-like domain. P-selectin binds to sialyl-Lewis^x on an O-linked glycan and a sulphated tyrosine residue both near the N-terminus of PSGL-1.

P-type carbohydrate recognition domain (CRD): One of the mannose 6-phosphate-binding domains found in the mannose 6-phosphate receptors.

Ramachandran plot: A plot showing the torsion angles φ and ψ for each amino acid residue in a protein.

Receptor for low-density lipoproteins: An endocytic receptor that mediates uptake of the cholesterol-containing low-density lipoprotein particles into cells.

Reducing end: The end of a disaccharide, polysaccharide, or oligosaccharide that retains the carbonyl function.

Rhizobial bacteria: A group of soil bacteria that fix nitrogen. They grow symbiotically in root nodules of leguminous plants.

Ricin: A highly toxic plant protein found in the seeds of the castor oil plant. It enters cells by endocytosis after binding to galactose residues on cell surface glycoproteins.

R-type CRD: Ricin-like carbohydrate-recognition domain. A carbohydrate-binding domain homologous to the galactose-binding domains of the plant toxin ricin.

Schwann cell: A cell found in the peripheral nervous system that produces the myelin sheath for coating the axons of neurons.

Selectin cell adhesion molecules: A group of three (L-selectin, P-selectin, and E-selectin) membrane-bound C-type lectins that mediate adhesion of leucocytes to endothelial cells.

Sialidase: An enzyme that cleaves sialic acids such as NeuAc from glycans. Commonly referred to as neuraminidase.

Sialoadhesin: A sialic-acid binding protein found on macrophages. It is the largest member of the siglec family of I-type lectins, containing 17 immunoglobulin-like domains.

Sialyl-Lewis^a: Neu5Acα2,3Galβ1,3(Fucα1,4)GlcNAc. A tetrasaccharide found as a terminal elaboration on both N-linked and O-linked glycans.

Sialyl-Lewis^x: Neu5Acα2,3Galβ1,4(Fucα1,3)GlcNAc. A tetrasaccharide found as a terminal elaboration on both N-linked and O-linked glycans.

Sialylmotifs: Short, conserved sequences characteristic of all sialyltransferases.

Sialyl-Tn epitope/antigen: An O-linked glycan NeuAcα2,6GalNAcα-Thr/Ser expressed at high levels on a variety of carcinomas.

Sialyltransferase: An enzyme that catalyses transfer of sialic acid from the activated donor CMP-sialic acid to an acceptor oligosaccharide during synthesis of glycoproteins or glycolipids.

Siglecs: A group of structurally related sialic acid-binding proteins with I-type carbohydrate-recognition domains. Examples include sialoadhesin, CD22 and myelin associated glyco-protein (MAG).

Site analysis: Determination of what glycans are attached to specific amino acid residues in a protein.

Sulphatide: A galactosphingolipid with galactose 3-sulphate linked to ceramide.

Surfactant proteins A and D (SP-A and SP-D): Two members of the collectin subgroup of C-type lectins found in the lung.

Swainsonine: A plant alkaloid that is a reversible inhibitor of Golgi mannosidase II.

Syndecan: One of a family of four transmembrane heparan sulphate proteoglycans found on a variety of cell types. The heparan sulphate chains of syndecan interact with growth factors at the cell surface.

Systems biology: An approach to analysing and integrating complex sets of data in biology using mathematical modelling to predict the behaviour of a complete system, such as a cell or organism, based on the properties of its component parts.

Tay–Sachs disease: A glycolipid storage disease caused by deficiency of the enzyme α-hexosaminidase A that catalyses degradation of ganglioside GM2.

Tetra-antennary: Having four branches. Refers to a type of N-linked complex oligosaccharide that has four branches of sugars attached to the trimannose core. Two branches extend from each arm of the core.

Tissue plasminogen activator: A serine protease enzyme made by cells lining blood vessels. It activates plasminogen which acts to dissolve blood clots.

Tri-antennary: Having three branches. Refers to a type of N-linked complex oligosaccharide that has three branches of sugars attached to the trimannose core. Two branches extend from the 1–6 arm of the core and one branch from the 1–3 arm.

Trypanosome: A single-celled protozoan of the genus *Trypanosoma*. They live as parasites in the bloodstream of vertebrates. Many species do not harm their hosts, but some cause serious diseases. Examples include *Trypanosoma cruzi*, which causes Chaga's disease, and *Trypanosoma brucei*, which causes sleeping sickness.

Tunicamycin: A bacterial compound that inhibits the reaction of UDP-GlcNAc and dolichol in the first step of synthesis of the precursor oligosaccharide in N-linked glycosylation. It is used in research to prevent N-linked glycosylation of proteins.

Type II transmembrane protein: A protein with a cytoplasmic N-terminus, a single transmembrane segment, and a C-terminal extracellular domain.

Ubiquitin: A 76-amino acid protein that becomes covalently attached to other proteins. The C-terminus of ubiquitin forms a bond to the amino group of a lysine residue of the target protein. Addition of a chain of ubiquitin molecules targets a protein for degradation in the proteasome.

Variant surface glycoprotein: A glycoprotein anchored to the membrane of a trypanosome by a glycosylphosphatidylinositol anchor. Variant surface glycoproteins form a dense protective coat at the trypanosome cell surface. Each parasite has genes for several hundred different variant surface glycoproteins.

Versican: A chondroitin sulphate proteoglycan expressed in many tissues. It contains 12–16 chondroitin sulphate chains.

VIP36 (Vesicular integral membrane protein of 36 kDa): An L-type lectin that binds high-mannose oligosaccharides of glycoproteins in the endoplasmic-reticulum-Golgi intermediate compartment (ERGIC).

VIPL: VIP36-Like protein, a member of the L-type lectin family.

von Willebrand factor deficiency: A blood clotting disorder caused by insufficient levels of von Willebrand factor.

Xenoantigen: An antigen expressed by one species but not by another species.

Xylosyltransferase: An enzyme that transfers xylose from the activated donor UDP-xylose to an acceptor oligosaccharide during synthesis of glycoproteins or to the core protein of proteoglycans.

Pages that include figures are highlighted in blue.